Genes and Human Self-Knowledge

Genes and Human Self-Knowledge

Historical and Philosophical Reflections on Modern Genetics

Edited by Robert F. Weir,
Susan C. Lawrence, and Evan Fales

University of Iowa Press Ψ Iowa City

University of Iowa Press, Iowa City 52242

Printed on acid-free paper

The symposium on genes and human self-knowledge was supported
by the Iowa Humanities Board, the National Endowment for the
Humanities, the University of Iowa Center for Advanced Studies,
Office of Academic Affairs, College of Medicine Lecture Committee,
University Lecture Committee, College of Law, Department of
Pediatrics, Department of History, School of Religion, Program in
Biomedical Ethics, History of Medicine Society, Humanities Society,
and the Iowa Medical Society. The descriptions, opinions, and
conclusions in this book do not necessarily represent the views of
the Iowa Humanities Board, the National Endowment for the
Humanities, or any of the other sponsors.

Library of Congress Cataloging-in-Publication Data
Genes and human self-knowledge: historical and philosophical
reflections on modern genetics / edited by Robert F. Weir,
Susan C. Lawrence, and Evan Fales.
p. cm.
Includes bibliographical references.
ISBN 0-87745-455-8, ISBN 0-87745-456-6 (pbk.)
1. Human Genome Project—Moral and ethical aspects.
2. Human genetics—History.   3. Human genetics—Philosophy.
I. Weir, Robert F.   II. Lawrence, Susan C.   III. Fales, Evan.
QH445.2.G443   1994   93-45453
573.2′1′01—dc20   CIP

98 97 96 95 94 C 5 4 3 2 1
98 97 96 95 94 P 5 4 3 2 1

Contents

Preface

Contemporary developments in the field of human genetics are profoundly important, in regard to both the rapidity of new scientific discoveries and their personal and social implications for all of us. Of course not all of the discoveries are of the magnitude of the mapping and sequencing of the gene on human chromosome 7 that causes cystic fibrosis or of the mapping and sequencing of the gene on chromosome 4 that causes Huntington disease. Nevertheless, hardly a day passes without media reports of yet another medical researcher or group of researchers who, in the role of medical detectives, have discovered another genetics clue in the long-running mystery of human identity. Related stories in the media frequently raise questions about the benefits and risks of new genetic information to individuals, families, and the general public.

To cite one example, a clinical genetics case in Michigan was given considerable attention by the media in January 1993. The case combined, first, the discovery by a team of molecular geneticists of the gene that makes women in some families susceptible to developing breast cancer; second, several women in an affected family who had been screened for the susceptibility gene and wanted to know the scientific information about themselves; and, lastly, a group of genetic counselors who had to decide if their scientific facts were sufficiently accurate to give the women the intensely personal information that they desperately wanted. The stakes were high: several women in the last three generations of the family had died from breast cancer, several others had chosen to have prophylactic bilateral mastectomies, and one woman confronted her genetic counselor with the options of telling her the genetic truth about herself or standing by as she proceeded to have her breasts surgically removed the following week. After deliberation with colleagues, the counselor told her that, based on the team's molecular research, there was a 98 percent chance that she did not carry the defect and thus had no more than the standard 10 percent risk of developing the disease. The woman cried, laughed, danced around the counselor's office, canceled the scheduled surgery, experienced liberation from the dread she had lived with all her adult life, and began to live with new self-knowledge—including survivor guilt as she compared her genetic status with that of some of the other women in her family. The genetic counselors, meanwhile, took steps to protect this new scientific and personal knowledge from getting into the wrong hands, such as insurance companies that might deny coverage for women in the family who had inherited the susceptibility gene.[1]

Several months before this publicized case, in April 1992, the University of Iowa sponsored a four-day symposium that brought together historians, philosophers, researchers in molecular genetics, genetic counselors, physicians, biomedical ethicists, students, and members of the general public to discuss multiple aspects of human genetics. The title of the 1992 University of Iowa Humanities Symposium was the same as the title for this book: "Genes and Human Self-Knowledge: Historical and Philosophical Reflections on Modern Genetics." Joining the participants from Iowa City were James Watson, at the time the director of the National Center for Human Genome Research, as the keynote speaker; a number of distinguished scholars from the United States and Canada in a variety of academic fields, who delivered the papers published in revised form in this book; and one student and several representative faculty from colleges and universities in Iowa who joined University of Iowa faculty as panelists with prepared responses to the invited papers presented at the symposium.

The purpose of the symposium was twofold. First, in regard to participants, the symposium was planned to maximize the benefits of serious, interdisciplinary discussion of issues important to all of us. To that end, geneticists, academic humanists, and members of the audience were invited to raise troubling questions, to probe unexamined assumptions and beliefs, to be open to new and different ideas, to listen, and to be candid about personal and professional disagreements.

Second, in regard to content, the symposium was planned so that some of the historical background, philosophical implications, and ethical issues related to the Human Genome Project (HGP) would be subjected to scholarly analysis, but not to the exclusion of other developments in modern genetics. To that end, many of the papers and much of the discussion at the symposium focused on the HGP, the $3 billion, fifteen-year scientific project in the United States (part of the international Human Genome Initiative) to map and sequence the estimated fifty thousand to one hundred thousand human genes that comprise the genetic blueprint for human beings.[2]

Projected to be completed in 2005, the HGP is a joint effort of the National Institutes of Health and the Department of Energy. The long-term benefits of the knowledge to be gained through the HGP are several: an increased awareness of how we function as healthy human beings, greater knowledge of the biochemical basis of many of the approximately four thousand human genetic diseases, intensified efforts to diagnose and treat genetic diseases, a more complete knowledge regarding the relationship of genetic and environmental influences on individual human development, and an increased understanding of how humans compare at a genetic level with the members of other species.

The overall goal of the HGP is to determine all of the information in the "basic" human genome, meaning the standard, predictable molecular genetic information contained in a typical composite of the "normal" human chromosomes: the twenty-two autosomal chromosomes, plus the X chromosome, the Y chromosome, and the mitochondrial chromosome. Although the literature on the HGP is replete with references to "the" human genome, the scientists working on the HGP clearly do not mean to suggest that there is any model, or ideal, or singular example of human genetic makeup. In fact, much of the scientific interest in the HGP has to do with variations from genetic normalcy found at the molecular level, including variations among individual humans as well as variations between "the" human genome and selected nonhuman genomes such as the bacterial, fruit fly, mouse, and yeast genomes.

To achieve this goal, the HGP has two interrelated objectives. The first objective is to develop detailed maps, or descriptive diagrams, at increasingly finer resolutions, of the human genome and other selected genomes. Such maps come in two types. Genetic linkage maps assign a distance between markers (identifiable fragments of DNA that can vary from individual to individual) on a chromosome according to the frequency with which they are inherited together, with more closely linked markers being more likely to be part of the same inheritance pattern. The linkage, or likelihood of inherited frequency, between markers is measured in units called centimorgans. Physical maps, by contrast, indicate the actual physical distance between markers without regard to inheritance. In these maps, the units of measurement are base pairs, or nucleotides, the chemical building blocks of the genome.

The second objective of the HGP is to determine the complete base sequences of the human genome, as well as selected nonhuman genomes, with various genomes differing significantly in terms of the total base pairs they contain (e.g., the human genome has 3 billion base pairs, whereas the *Escherichia coli* genome has only 4.6 million base pairs). To do so requires the improvement of existing sequencing technologies (in terms of efficiency and cost per sequenced base pair), the discovery of new sequencing technologies, and the development of new systems of informatics to handle the storage and communication of increasingly sophisticated and detailed scientific information.[3]

Among the notable features of the HGP is the policy decision to allocate 5 percent of the funding to work done on the ethical, legal, and social implications of the project, as illustrated by some of the essays in this volume.[4] By contrast, some of the other essays hardly mention the HGP, with the authors preferring to provide a historical interpretation of earlier developments in genetics or a philosophical analysis of fundamental questions and ethical issues in genetics that are related to but not limited to the HGP.

The symposium was organized around a number of questions pertaining to the implications, benefits, risks, and problems brought about by modern genetics, including the HGP. Some of the questions are epistemological in nature: What kind of knowledge is being discovered or produced by molecular geneticists? When James Watson and other proponents of the HGP make comments like "our fate is not in our stars but in our genes" and "we will now really know ourselves at the molecular level," what kind of epistemological claims are they making? Is human self-knowledge at the molecular level different in important ways from human self-knowledge gained in other ways? Is knowledge of self (and others) gained through the technologies of molecular biologists more fundamental than knowledge of self (and others) gained through psychological counseling, anthropological or sociological research, or historical studies?

Other questions focus on the implications of new genetic knowledge in terms of identifying who we are in relation to others: Is "the" human genome largely homologous with the genomes of other species? Do individual human genomes differ significantly from one another? Do most males and most females differ from each other at the molecular level, as well as in terms of one of the chromosomes they inherited? Is there any scientific basis at the molecular level for distinguishing among racial or ethnic groups?

Additional questions pertain to the historical uses and abuses of genetic knowledge, as well as to possible future uses and abuses of this knowledge: Is the contemporary use of scientific information in genetic counseling a modern parallel to the eugenics movement in the early part of this century? Will current work in molecular genetics lead to updated views of genetic "normalcy" or genetic "superiority," as occurred in the United States and Germany in the first half of this century? Will knowledge of our genetic identities at the molecular level be used by employers, insurance companies, the military, and the government for the purposes of stigmatization, discrimination, and loss of employment or insurance? Should medical researchers and physicians intervene in someone's genetic makeup to prevent or cure genetic disease, either by means of somatic-cell gene therapy or germ-line therapy? Should the scientific knowledge and skills to change an individual's genome be used to enhance that person's capacities as well as to prevent or cure genetic disease?

Still other questions concern the practical problems of communicating molecular biology to nonscientists: How can members of the print and electronic media best convey the basic features of a highly specialized, esoteric knowledge to the general public? Can metaphors, analogies, artistic drawings, and other visual aids communicate the scientific information that individuals need to know about themselves and their families? Is it possible to "translate the laboratory" so that nonscientists can understand, at least at a fairly simple level, the work being done by molecular geneticists? What steps are being taken, and what additional steps need to be taken, to help high school students become scientifically literate?

Finally, some questions focus on the personal and private aspects of genetic knowledge: What kind of self-knowledge do counselees gain through the process of genetic counseling? Does gaining new genetic information increase one's self-understanding? When persons in genetic counseling find out new information about their genetic makeup, their genetic heritage, and the impact of their genetic makeup on their children, what kind of self-knowledge do they now have that they did not possess earlier? Do most people want to be given this kind of self-knowledge if, as in Huntington disease, it is potentially devastating to them and the future they had planned? How do and how should counselees respond to the unwelcome knowledge that they have contributed to or caused the genetic abnormality their daughter or son now has? Are some persons better off without this kind of self-knowledge gained through genetic counseling?

Many of these questions were addressed in the symposium papers that are published in revised essay form in this book. All of the essays, with two exceptions, are original to this volume. They are organized in three thematic sections, with the major papers delivered at the symposium being followed by selected shorter responses presented by panelists.

Section I, entitled "Genetic Identity and Self-Knowledge," contains three major essays. Kimberly Quaid, a psychologist, observes that genetic information can have a significant impact on one's self-knowledge, one's sense of freedom and responsibility, and one's self-esteem, as indicated by a series of personal statements by individuals at risk for Huntington disease. Dan Brock suggests in a philosophical analysis that the HGP may influence our understanding of human identity in three fundamental ways: our views of equality, our sense of personal responsibility for conduct, and our standards of normality. Michael Ruse, also a philosopher, cautions that the work of molecular geneticists may add weight to various theories of reductionism, at great cost to our conceptions of human freedom, responsibility, and variability.

The first section includes the responses of five panelists. Panayot Butchvarov, William Carroll, and Evan Fales address the comments made by Brock and Ruse regarding reductionism, determinism, and free will, with each of these panelists raising questions regarding the degree to which the HGP may actually influence the debate over these perennial philosophical issues. Diana Cates follows with a commentary on the relationship between genetic knowledge and one's self-understanding. Christine Carney, a student at the University of Northern Iowa, then shares her very personal views as a consumer of genetic services, discussing how osteogenesis imperfecta has influenced her life, her family, and her plans regarding procreation.

Section II has four major essays that address, in quite different ways, possible uses and abuses of genetic knowledge. Diane Paul, a historian of science, challenges us to consider the possibility that we may have entered a new era in which genetic knowledge will be abused through the practices of "back-door" eugenics, characterized by parental demands for genetically perfect babies. Joseph McInerney, a biological sciences educator, focuses on the uses of genetic knowledge in education, arguing that high school students urgently need to learn about molecular biology to help them have a better understanding of themselves, their genetic inheritance, and the world in which they live. Larry Thompson, a journalist, describes some of the ways in which genetic knowledge is both used and abused by reporters, editors, and others in the media as they try to communicate important information to a largely nonscientific audience. Larry Gostin, a legal scholar, analyzes the Americans with Disabilities Act of 1990 and warns that additional legal safeguards are needed against genetic discrimination, lest many persons be denied employment and/or insurance based on genetic conditions they have, carry, or might acquire because of the interaction of their genetic makeup with certain environmental factors.

This section has short essays from six panelists. Mitchell Ash and Alan Marcus both respond specifically to Paul's essay, and both agree with her that eugenical thinking, at least, and possibly eugenical practices remain a characteristic feature of this century, more so in some countries than in others. Elizabeth Thomson and Kevin Koepnick address some of the uses and abuses of genetic knowledge, with both of them pointing out some of the serious problems of communicating specialized scientific information to persons lacking significant scientific education, whether in genetic counseling situations or in high school biology classrooms. The responses by Peter Blanck and Robert Weir conclude the section, with both of them addressing some of the abuses that can and do occur when genetic knowledge about individuals and families is obtained by third parties.

Section III contains three major essays that address various aspects of the theme "Genders, Races, and Future Generations." Ruth Hubbard, a biologist, maintains that ideological perspectives, not biological facts, provide the basis for most of society's constructs of racial and sexual differences. David Hull, a philosopher, uses the perspective of evolutionary biology to emphasize the importance of genetic variation among species, genders, and races, and to argue that variation is not the same thing as deviation. LeRoy Walters, a biomedical ethicist, addresses the issue of genetic interventions that may affect future generations and concludes that, in at least some instances, human germ-line genetic interventions can be justified for the cure or prevention of genetic disease.

The third section concludes with short essays written by a historian, a philosopher, and a religious ethicist. Susan Lawrence responds to Hubbard and Hull by emphasizing the importance of accuracy and accountability in the language used by scientists and humanists. David Magnus highlights an important difference between the views of Hubbard and Hull and expresses concern regarding the reductive approach of the HGP, a concern shared by other persons at the symposium. John Boyle, in his response to the essay by Walters, expresses a different concern: the possibility of long-lasting harm to future generations that could result from germ-line genetic interventions.

The symposium on genes and human self-knowledge was an important event that had influence far beyond the campus of the University of Iowa. We hope that the publication of the papers from the symposium will also be an important contribution by academic humanists and other professionals to the interdisciplinary literature on modern genetics, including the HGP.

R. F. W.

NOTES

1. Leslie Roberts, "Genetic Counseling: A Preview of What's in Store," *Science* 259 (January 29, 1993): 624.
2. The literature on the Human Genome Project is expanding rapidly. Books on the subject include Jerry Bishop, *Genome* (New York: Simon and Schuster, 1990); Daniel Kevles and Leroy Hood, eds., *The Code of Codes: Scientific and Social Issues in the Human Genome Project* (Cambridge, Mass.: Harvard University Press, 1991); Bernard D. Davis, ed., *The Genetic Revolution* (Baltimore: Johns Hopkins University Press, 1991); Thomas Lee, *The Human Genome Project* (New York: Plenum, 1991); Robert Shapiro, *The Hu-*

man Blueprint (New York: St. Martin's Press, 1991); Christopher Wills, *Exons, Introns, and Talking Genes* (New York: Basic Books, 1990); David A. Micklos and Greg A. Freyer, *DNA Science: A First Course in Recombinant DNA Technology* (New York: Cold Spring Harbor Laboratory Press, 1990); George Annas and Sherman Elias, eds., *Gene Mapping: Using Law and Ethics As Guides* (New York: Oxford University Press, 1992); David Suzuki and Peter Knudtson, *Genethics* (Cambridge, Mass.: Harvard University Press, 1990); Anthony Holtzman, *Proceed with Caution* (Baltimore: Johns Hopkins University Press, 1989); and Theodore Friedmann, ed., *Molecular Genetic Medicine* (San Diego: Academic Press, 1991).

3. See James D. Watson, "The Human Genome Initiative: A Statement of Need," *Hospital Practice* 26 (October 15, 1991): 69–73; Office of Health and Environmental Research, U.S. Department of Energy, *Human Genome: 1991–92 Program Report* (Washington, D.C.: U.S. Department of Energy, 1992), pp. 192–218; and Biological Sciences Curriculum Study (BSCS) and the American Medical Association, *Mapping and Sequencing the Human Genome: Science, Ethics, and Public Policy* (Colorado Springs, Colo.: BSCS, 1992), pp. 1–3.

4. The literature specifically addressing ELSI issues is also expanding rapidly. See U.S. Department of Energy, *Bibliography: Ethical, Legal, and Social Implications of the Human Genome Project* (Washington, D.C.: U.S. Department of Energy, 1993), compiled by Michael Yesley, coordinator of the ELSI Program at the Department of Energy.

Acknowledgments

The 1992 University of Iowa Humanities Symposium was an interdisciplinary conference whose theme, "Genes and Human Self-Knowledge," attracted faculty from many university departments, students, faculty from other colleges and universities in Iowa, professionals from other states, and members of the general public. The symposium was planned in such a way that it would help "bridge" the Iowa River that flows through the campus, providing a public forum for an exchange of ideas and perspectives between molecular and clinical geneticists (most of whom work on one side of the river) and academic humanists (most of whom work on the other side of the river).

Planning the symposium involved many faculty members. Mitchell Ash, from the Department of History, first proposed that we focus the symposium on the subject of modern genetics. Susan Lawrence, from the College of Medicine and the Department of History, took responsibility for publicity. Evan Fales, from the Department of Philosophy, and Jay Semel, director of the Center for Advanced Studies, helped with the arrangements for invited speakers and panelists. Jeffrey Murray, from the Department of Pediatrics, planned three pre-symposium workshops on genetics for persons outside the biological sciences. John Boyle, from the School of Religion, helped manage the budget. Other members of the planning committee were Panayot Butchvarov (Philosophy), Gary Gussin (Biology), James Hanson (Pediatrics), and John Menninger (Biology).

The symposium was actively supported by the university administration. Hunter Rawlings, the president of the University of Iowa, provided a welcome to the opening session of the symposium by communicating his ongoing, strong backing for educational and research efforts that transcend the walls that sometimes separate departments and colleges in a university. Peter Nathan, the vice president for Academic Affairs, played a key role in the symposium not only by moderating one of the sessions, but also by revealing how a serious genetic condition has deeply affected his family. James Clifton, the interim dean of the College of Medicine, and William Hines, dean of the College of Law, also helped support the symposium in significant ways.

Funding for the symposium came from several sources. Major grants were received from the University of Iowa Office of Academic Affairs, the Iowa Humanities Board, and the National Endowment for the Humanities. Additional funding came from the College of Medicine, the College of Law, the Iowa Medical Society, and a number of departments and programs throughout the university.

Three secretaries helped in the preparation of the manuscript. To Brenda Burdick, Renée Choe-Winter, and Melanie DeVore, my thanks for your patience, interest in the project, helpful suggestions, and attention to details.

At the University of Iowa Press, Paul Zimmer, Holly Carver, and their associates have been enthusiastic about publishing the symposium papers, patient with us as we have wrestled with unexpected problems, and committed to turning out a first-rate book. They hope, and we hope, that the papers included in this volume will prove to be valuable contributions to the literature on the personal and social impacts of human genetics in the last decade of this century.

R. F. W.

I. Genetic Identity
and Self-Knowledge

A Few Words from a "Wise" Woman

Kimberly A. Quaid

In his book *Stigma,* the sociologist Erving Goffman considers two sets of individuals from whom a stigmatized person can expect some support. The first set consists of those who share the stigma and by virtue of this are defined, and define themselves, as his or her own kind. The second set is described, borrowing a term used by homosexuals, as the "wise," namely, persons who are "normal" but whose special situation has made them intimately privy to the secret life of the stigmatized individual and sympathetic with it, and who find themselves not only accepted but often offered a courtesy membership in the clan.

As Goffman put it:

A discrepancy may exist between an individual's virtual and actual identity. This discrepancy, when known about or apparent, spoils his social identity; it has the effect of cutting him off from society and from himself so that he stands a discredited person facing an unaccepting world. In some cases, he may continue through life to find that he is the only one of his kind and that all the world is against him. In most cases, however, he will find that there are sympathetic others who are ready to adopt his standpoint in the world and to share with him the feeling that he is human and essentially normal in spite of appearances and in spite of his own self-doubts.[1]

Five years of experience counseling individuals seeking presymptomatic testing for Huntington disease has led me to an understanding of the secret life of persons who are at risk for this devastating disorder and of those who have discovered through genetic testing whether they carry the gene for this disease and may, or may not, be destined to develop the symptoms at some point in the future.

Huntington disease (HD) is a late-onset autosomal dominant genetic disorder with complete penetrance. That is, if one of your parents is affected with the disease, you yourself have a 50 percent chance of carrying the gene. If you carry the gene you will, barring death from other circumstances, develop the characteristic symptoms of the disease at some time in your life. These symptoms include personality and mood changes, primarily depression, cognitive changes including dementia, and disturbances in both voluntary and involuntary movements.[2]

The average age of onset is thirty-eight, although the disease can occur as young as two or as old as seventy.[3] Those affected can live on the average of sixteen years after diagnosis, but in many cases much of that time is spent in a nursing home in almost complete dependence. There is, at this time, no cure, and treatments are only palliative at best.

Huntington disease was the first disease mapped to a previously unknown genetic location on chromosome 4 using the technique of restriction fragment length polymorphisms (RFLPs).[4] The discovery of genetic markers linked to the HD gene, markers that traveled with the gene during reproduction, meant that some people at risk for HD could now learn whether they carried the gene.

This discovery has been a mixed blessing. From the beginning the possibility of freeing those at risk from the constant worry that every bobbled glass or forgotten telephone number meant the beginning of an inevitable decline needed to be balanced against the fact that some at-risk individuals would find out that they carried the HD gene and would have to live with this information for many years before their symptoms began. Surveys conducted before the test became available suggested that a certain percentage of those who learned that they carried the gene would consider suicide, thus raising concern about the effects of this information on those with a high risk.[5]

Predictive testing for HD was first offered on a research basis in order to gather information about the effects of presymptomatic testing. In the fall of 1986, two centers in the United States, funded by the National Institutes of Health (NIH) at Massachusetts General Hospital in Boston and at Johns Hopkins Hospital in Baltimore, began offering presymptomatic testing. Both of these institutions had well-established HD clinics and were well versed both in the manifestations of the disease and in dealing with families at risk for HD. These pilot programs were designed to assess the psychological and social ramifications of performing this type of testing. The research protocols included neurological screening, psychiatric screening, psychological assessment, pre-test counseling, and post-test follow-up. Preliminary results from these projects suggested that this testing could be done safely in a supportive setting with appropriate long-term monitoring of those tested.[6] By the end of 1991, twenty-three centers in the United States were offering presymptomatic testing for HD. The majority of those centers had opened within the previous year, and most had completed fewer than five such tests.[7]

To date, approximately 225 individuals in the United States have completed testing out of an estimated 125,000 at risk.[8] Although survey studies completed before the advent of testing indicated that high percentages of those at risk would want to be tested,[9] the numbers of those actually coming forward to be tested range from 9 percent to 16 percent of those eligible in areas where testing was offered free of charge.[10]

Follow-up studies of those who have been tested suggest that changes in self-perception can make coping with test results more difficult both for individuals whose test results indicate high risk[11] and for those whose test results indicate low risk.[12] In order to explore this issue in more depth, a notice was placed in *The Marker*, the newsletter published by the Huntington Disease Society of America (HDSA), requesting at-risk individuals to share their thoughts and feelings about being at risk for a serious genetic disorder, about how the introduction of genetic testing has affected these thoughts and feelings, and about the ways in which receiving test results has affected their self-image. This approach is based on the notion that stories are essential as a means of perceiving how scientific knowledge, in this instance knowledge of one's own genetic makeup, affects individuals.[13] Wherever possible, the words of those at risk for this disorder will be used in this essay.

An underlying theme in many of these stories is the special nature of genetic information and predictive testing. Information that one has, or is at risk for, a genetic condition is more intensely personal than information about an illness contracted as a result of contact with an external cause, such as a virus. Genetic information is widely viewed as saying something about who the person is at some fundamental, if unarticulated, level. In that sense, especially, people appear to feel stigmatized by exposure of their genetic information, and others might be more likely to stigmatize on that basis.[14] In some cases, even medical professionals exhibit the tendency to treat proven genetic disorders in a manner different from the way they might treat other diseases, with predictable effects on individuals at risk. Genetic information is also different from other types of medical information in that it involves other persons. An accurate diagnosis of a genetic disease in one member of a family is likely to alter the risks to several individuals both vertically (i.e., parents and children) and horizontally (i.e., siblings) throughout a family. Finally, predictive genetic testing is different from other medical tests. Unlike most medical tests, predictive genetic tests try to resolve more than the uncertainties of an existing condition by providing a specific diagnosis, or by directing an appropriate choice of treatment modalities. Predictive tests aim to resolve the uncertainty that is characteristic of the human condition.[15]

These aspects of genetic information are well known to people who have a genetic disorder running through their family. Most individuals at risk for an autosomal dominant genetic disorder have firsthand knowledge of both the disease and its effects. Many at-risk individuals have had to contend with years of secrecy and denial on the part of family members who refuse to believe that there is anything wrong. People talk in whispers about their shame at having this "taint" in their family. In one small town, HD came to be known as Cooke's disease after the family who had many affected members. One young woman at risk told me that she couldn't wait to be old enough to leave for college so that she could be anonymous. She knew many boys in town whose parents wouldn't let their sons date her because her own mother had "that disease."

But what is so private about genetic information is, at the same time, extremely communal. Genes are shared with ancestors, siblings, and children. In many cases, a person cannot even find out about his or her own genetic heritage without having information about numerous other family members. Genetic diseases are truly a family affair, with all the attendant problems of family interactions. The most eloquent description of this state of affairs came from a thirty-year-old woman, married, with two children, and at risk for HD.

My history, as far back as I know, begins with my great-grand-mother. I believe back then it was thought to be that the person who became like this was described in a hushed voice as "crazy." I'm not sure how she passed away. Next, my grandfather and his brother became victims of HD. My great-uncle was thought to be a drunkard and thrown in jail. There he was beaten to death by the guards who thought him to be a "wasted-minded" person due to all that "alcohol." Imagine, picked up off the street and murdered. It makes me cry, and my heart breaks for a man I never even knew. Then, my grandfather started to show the symptoms that his mother and brother had had. He shot himself in the basement when my mother was twelve years old. He left behind a wife, three sons, and a daughter. Elderly relatives have said he be-came very depressed when he lost his job as an engineer for a railroad company. He could no longer work due to this "shaki-ness" and "tremors." This crushed him. This gentle, loving man had suddenly turned into a violent man, and then into a very de-pressed soul.

When my sister and I were ten years old, my mother started slurping a lot and left the house a mess (she had been very particu-lar in her housecleaning duties, and it was often said that you could eat off my mother's floor, even in the basement). She started spending a lot of time in the hospital. It took months, but she was finally diagnosed with HD. I was only ten, but these words meant something terrible as my poor dad said them. This crushed my fa-ther. He had worshipped this woman at one time, now it was changing rapidly. My mother became very violent. My father was the main target of these what can only be described as rages. She destroyed him. He did everything for her and us. He got laid off, and then I really pitied him. He got to take her wrath twenty-four hours a day, seven days a week. She became violent more often, the good moods were becoming scarce. I came to hate her. A part of me still does, God forgive me. She got good at being almost a devil woman when it was just family—to a woman who you would swear clapped a halo on her head and sprouted wings in front of company. Even at age ten or eleven, I knew she was losing it. I dreaded her doctors' appointments. They would try different drugs on her. I would like them to know someday what even more of a nightmarish hell that was in itself. Sometimes, depending on what drugs she was on, she'd sleep all day and be wide awake all night to mentally and physically torture us. To this day I have to be the last one to bed at night because I still get this old feeling that something will happen. Subconsciously, I think I'll wake up from a slap across the head or the face. My father drank before this all happened, and now he started to drink more. Eventually he went over the edge. It was like she dared him to do what he later did. No matter what he did it wasn't enough. He was no good to her, to us, or anyone, I heard her say many times. Well, one Saturday evening (no drinking was involved) after one of her many tirades, he, like her father, shot himself. I was twelve years old. My brother and sister and I were devastated. She, I can still remember her face, sat there after the ambulance took him away, sat there with a smirk on her face. Later on, she would tell me and my brother we were better off without him. I had grown to dislike her over the long months, but at that moment, I realized how much I really hated her. And things were going to be worse over the coming years. My sister and I carried most of the burden, because we were "the girls." Our brother had gotten married and moved away. My other brother was living with a girlfriend, and our youngest brothers were sent to a home for the boys of veterans. There we were. I became a terrible teenager, drugs, booze, whatever. I only went out after I got home from school, made supper, did the dishes, laundry, and cleaning, and bathed Mom, gave her her pills, and put her to bed. Then I would sneak out. Eventually I got pregnant and moved out. My great-aunt came to help for a while, but ten months later my mother went into a nursing home. My sister visited maybe five times in the eight years she was there. Me—the one she hated—I tried to visit every other week. Part out of guilt, more out of duty. She died when I was twenty-six.

Her brother, my uncle, who had been diagnosed three years before she died, died two and a half years later. Right before my mom died, my oldest brother was diagnosed with the disease. Soon after, a team came from the medical center and examined us all, including my other brother. He had it, we knew it, but wouldn't say it to him for fear of what he would do. Our fears were realized five months later. He shot himself. One more blow. I am still reeling from that. We all are. This disease has devastated us. We have fallen apart. We have suffered individually, and this has put great distance between us. In times such as these, you should all pull together, but I am sure that nine times out of ten it doesn't happen that way. HD to me is like that phrase "comes like a thief in the night." It comes and steals your family—steals the love, steals the happy times. It leaves only the emptiness—the never-heard voices in the walls of a family home that should have echoed with laughter and joy. When I go home sometimes, I get up in the middle of the night and sit in a chair in the living room. All is quiet. Instead of remembering and picturing the happy times, I remember the muffled crying, the oh-so-loud screaming and yelling. It's dark at night and my memories make it become even darker. I've married and moved away, bought a new house, and still my memories follow me. The HD threat is always looming.[16]

What can be most disturbing about genetic information is that the risks of finding out that you are a gene carrier for a specific disease may be not only psychological, that is, anxiety, depression, or family discord, but social.[17] These social risks can include desertion, stigmatization, discrimination, and potential loss of insurance or employment.[18] These are terrible risks for currently healthy individuals to run for the sake of self-knowledge, no matter how important that information might be either for themselves or for other family members.

Unfortunately, the attitudes of many health professionals toward genetic diseases can cause individuals at risk to internalize feelings of being "tainted." Primary physicians sometimes convey an aura of hopelessness surrounding any condition considered to be genetic. The effects of this attitude are described by a young woman at risk who finally got the courage to make an appointment with a neurologist after her father was diagnosed with HD. She writes:

I sat in horror as he relayed to me the details of the disease. He painted a bleak, hopeless picture and told me that I should consider having my tubes tied to avoid having children. He did not bother to refer me to any self-help groups, genetic counseling, or the Huntington Disease Society of America. The doctor left me feeling hopeless and of little value to myself or others. I was certain that I was damaged and defective and that no one would ever love me. Looking back, it is not surprising that I sank into a depression which became progressively deeper from that day on, culminating several years later in two attempts to take my own life.

A few years later, I discovered the National Hereditary Disease Foundation and became an occasional member of a support group. I began to read about advances in research theories and about drug treatment and behavior management. Although I still felt depressed, I began to gain some sense of control and power through knowledge and activity. At about that same time, I was invited to join my parents in a genetic counseling session. After great initial resistance, I agreed to participate. The doctor and counselor in the session expressed their confidence in my knowledge of my risk and counseled me in my right to self-determination. In this meeting, I felt I was given permission to live, I felt empowered and more hopeful than I had in some time.

The inherent threat in the availability of these tests is that the initial option of being tested may be transformed into a duty.[19] The issue is whether we, as a society, are willing to pay for genetic testing and screening programs in order to assist prospective patients and parents to make autonomous informed choices about their own lives, whatever those choices might be, or whether this information is to be provided only as a precursor to acceptable actions.

For those at risk, the answer seems clear. This same young woman continues:

At that time, I was also counseled about the newly developed predictive test. I decided against it for the following reasons: I decided that since I was in my midtwenties, there was too large a gap between knowing and onset. I was not certain I could lead a full or productive life with the knowledge that I would develop HD. I decided that the test would be more valuable to me at a later time if I were showing symptoms or if the gap between knowing and onset were smaller.

Although I wished to have children, I decided that I would probably still wish to have children regardless of the outcome of the test. I decided that life was good, and that even a shortened life span with diminished capabilities had value. I decided that persons at risk for other diseases—cancer, heart disease, diabetes—are not generally discouraged by the medical establishment, nor by society, from having children, and neither should I be so discouraged.

There is no treatment or cure for HD. What good would it do me to know now? There was nothing I could do to change the inevitable one way or the other. Would I really modify my behavior or lead my life any differently? A yes answer to that question would surely nullify the meaning of my present life. I decided that living my life to the fullest, with hope for the future, was the best possible solution.

My personal philosophy has not been one generally accepted by the majority of medical practitioners that I have known in the past ten years. I have been told that I live in denial because I do not wish to have the test. I have even been told that I am ignorant because I wish to have children! One gynecologist even called me at home without invitation after a routine office visit to try to advise me about alternatives to having my own children. I have begun to feel that these doctors believe that with enough genetic counseling I will eventually make the "right" decision in my life (that is, not to have children).

The use of genetic tests by professionals who have neither personal experience with the disease in question nor knowledge of families with these diseases increases the chances that an individual at risk for a genetic disease will be seen as a representative of a disease category and not as an individual. This trend exacerbates the tendency to place more emphasis on biology than on the personal, social, and environmental factors that may profoundly influence both the manifestation of symptoms and their meaning. A two- or three-paragraph description of a condition in a medical genetics text does not provide sufficient information either to understand the range of symptoms or to give advice on creative coping. One young man described the experience this way:

> When I heard my father had been diagnosed with HD, I knew nothing about it. I went to the medical library at school and looked it up. I sat on the floor in front of the stacks absolutely stunned when I realized that I, too, might have this. I have since thought many times that if a science fiction writer were to invent the worst disease in the world, he or she couldn't do a better job than this Huntington disease.

But the bleak descriptions and grainy black-and-white photographs found in medical textbooks do not begin to convey the courage of these families or their determination. Individuals faced with the choice whether to be tested for HD may be considered pioneers. They are setting off on a long journey with no maps. It is a journey fraught with danger due to the possibility of major decisions made about their lives and their selves based on fear and ignorance. We all may have to face similar versions of this journey.

One of my clients is a nursing student. She told me that one day in class they were talking about people in nursing homes. Her brother is in a nursing home in the later stages of HD. A discussion began concerning the uselessness of keeping people in nursing homes alive year after year. Unable to keep still any longer, my client told the class about her brother, and about HD, and about her own risk. The very next day, the two nursing students in her carpool failed to pick her up at the usual time. When called to explain, they voiced vague excuses but made it clear that they no longer cared to include her in their carpool and that she would have to make other arrangements.

Despite such reactions, individuals and families do learn to cope and to devise refreshing solutions to their plight. Three sisters who had recently been told of their risk came to learn more about HD and to consider testing. One day all three came in laughing and told me that they had it all figured out. Even if they all tested positive, they would pool their money and buy a diner. I asked how that would solve their problems, as I pictured lots of broken crockery and smashed glassware. "Oh," they replied, "we'd have a very limited menu. The only two things we would serve would be scrambled eggs and milk shakes."

These three women were actively working to imagine a future that might include the possibility of illness, dependence, and a shortened life span. Many of those at risk find the contemplation of their future a more difficult task. On another occasion I was able to tell a woman in her early thirties that she was at very low risk for carrying the HD gene. She said, "Thank you," and I said, "You're welcome, it's my pleasure." But she put her hand on my arm and said urgently, "No, you don't understand, you have just given me my life back." Several months later I had the chance to talk with her when she returned for her scheduled follow-up appointment. I asked her what, if anything, had changed. "Not much," she said. "Oh, I've decided to return to school, but that isn't the biggest thing. I'll tell you, the moment I knew that something had really changed was when I found myself making New Year's resolutions for the first time in ten years. It struck me then that I did, indeed, have a future."

Negative, or low-risk, results are not without psychological consequences. Follow-up studies of individuals who have been tested find that 10 percent of those with low-risk results have some difficulty in adjusting and require supportive counseling above and beyond the counseling provided by the testing protocol.[20] One important reason for a difficult adjustment in the face of what most people would consider unequivocal good news is that finding that you yourself are not a gene carrier does not mean that you never have to worry about this disease again. Rather, most of those tested have a parent who is currently ill and is likely to get worse, and most have at least one sibling who is either showing symptoms or is likely to show symptoms at some point in the future.

Those who have lived their lives at risk may find that their at-risk status has become a permanent part of their personality. One woman returned several months after receiving a low-risk result and revealed that during the time of the counseling and testing, she had made several bargains with herself along the lines of "if only I don't have HD, I will be sweeter, kinder, more patient, more tolerant of my family and friends, and more giving of my personal time and energy." "But," she said, obviously distressed, "I'm still such a bitch."

Low-risk results can also challenge assumptions about what an individual can expect from life. One thirty-seven-year-old woman had married a man with a different genetic condition. After receiving the news that she was at very low risk for carrying the HD gene, she decided to divorce him. She described her feelings in the following manner:

I used to feel that I was damaged and that people had some basis for discrediting me, that I may not have been fully intact because of the effects of HD, and that my emotional turmoil was not only psychological but biological in origin. When I found out that I have only a 3 percent risk factor, my vision of what my life held in store changed for me. I could be confident in having a "normal" life, a normal life span, and normal everything I may want. I changed my goal from being a career-oriented woman to that of remarrying a man who is healthy emotionally and physically, with whom I can have a child before I get too much older. I want to work part-time and care for my family, home, and new life. I hid behind my career so that I would not have to face being childless and financially dependent on a man, a man who would probably leave me when I became ill, as so many men do when HD manifests itself in a wife. I needed my own secure source of income. I could not trust that any man would be there if I became ill with HD.

Now I feel that I can get and have anything I want in relationships, that people see and treat me as a whole person without mental deficits. I fought to find out what my genetic material had to say and I am grateful that I am able to have a full life now. I am not angry anymore.

Those who have tested positive, or who have an increased risk, face a different set of challenges. One woman whose results indicated an increased risk wrote the following:

Louisa May Alcott said: "Far away, there in the sunshine, are my highest aspirations. I may not reach them, but I can look up and see their beauty, believe in them and try to follow where they lead."

I was once called a "Pollyanna" by an employer who didn't want to give the people under me a raise. He said, "Life isn't fair, don't expect it to be." I know firsthand that life can hand you a lot of problems and troubles that we really don't deserve. But, we cannot make these go away by denial or pretending they don't exist.

I guess the first few months I couldn't get past the horror of what was to become of me in the future. I struggled with it, I cried, I got angry. All of the normal feelings and emotions; I hurt.

At this time, my best friend told me the sparkle was gone from my eyes. She could tell how deeply I was hurting.

After several months, I felt that I needed to get back to life, get on with my life and live again.

I made a conscious effort to do just that. All that had happened to me in the past, I relived to see just how much I had accomplished, and how far I have come through some very rough times.

If I had been told I had cancer, I would have reacted the same way. I felt that I needed to hit it head on. It was now a fact of my life that I had Huntington's. How I choose to live from this point on is my choice. I can become bitter and hard, or I can get on with my life and change any long-term goals to shorter goals. I can readjust my wants and priorities. I needed to see where exactly I want to go over the next ten years.

I still have some thinking to do on those long-term goals.

On the short term, I know that my relationships are my top goal. I need to work to strengthen my family relationships, to keep them in good order. To show my family how much they mean to me.

I want to see my youngest son get through school and earn his law degree. I plan to help and work with him to reach that goal.

I have contact with persons dealing with life-ending cancer and heart problems. They would gladly exchange with me the time they have with what I have, even with the problems that go with it. I cannot see where being a negative and bitter person would be a benefit to me or the people around me.

My faith is very important to me. I believe that God is there for me all the time. Even when it gets rough, I can call on him to see that I will get through. There is no doubt that as a Christian I know where I will spend eternity. I hope that what I have to go through here on earth won't be more than I can handle. I do see that at times I feel like a "Pollyanna." . . . I feel that everyone should be treated fairly, treated well, and respected. I don't see me changing my outlook on that at all. I feel that as a person with Huntington's, I need to go on and be a positive person. Here is an-other quote that I especially like, by Harold Melchert: "Live your life each day as you would climb a mountain. An occasional glance toward the summit keeps the goal in mind, but many beautiful scenes are to be observed from each new vantage point. Climb slowly, steadily, enjoying each passing moment; and the view from the summit will serve as a fitting climax for the journey."

Many find the difficulty of defining their status unsettling. As one woman said, "I go to the support group meetings and we always go around the room and everyone introduces themselves and says whether they are affected, or at risk, or a spouse, or a caregiver. I'm not sure what to say anymore." I told her that if she wanted, we could make up a whole new category just for her. She said, "I don't really want a new category, I just want to be able to fit into an old category." But she can't. These people have become a new category, not only to themselves, but to society. We need to think about how to handle that category, even as these people are struggling individually to incorporate this new information into their view of themselves.

ACKNOWLEDGMENTS

My deepest gratitude goes to the many individuals at risk for Huntington disease who took the time and energy to write or call and to share with me their thoughts, some of which are expressed in this paper. Your courage and spirit humble me.

NOTES

1. E. Goffman, *Stigma: Notes on the Management of Spoiled Identity* (New York: Simon and Schuster, 1963).

2. S. E. Folstein, *Huntington's Disease: A Disorder of Families* (Baltimore: Johns Hopkins University Press, 1989).

3. L. A. Farrar and P. M. Conneally, "A Genetic Model for Age at Onset in Huntington's Disease," *American Journal Human Genetics* 37 (1985): 350–357.

4. J. F. Gusella, N. S. Wexler, P. M. Conneally, S. L. Naylor, M. A. Anderson, R. E. Tanzi, P. C. Watkins, M. R. Wallace, A. Y. Sakaguchi, A. B. Young, I. Shoulson, E. Bonilla, and J. B. Martin, "A Polymorphic DNA Marker Genetically Linked to Huntington's Disease," *Nature* 306 (1983): 234–238.

5. S. Kessler, T. Field, L. Worth, and H. Mosbarger, "Attitudes of Persons at Risk for Huntington's Disease toward Predictive Testing," *American Journal Med. Genetics* 26 (1987): 259–270; C. Mastromauro, R. H. Myers, and B. Berkman, "Attitudes towards Presymptomatic Testing in Huntington's Disease," *American Journal Med. Genetics* 26 (1987): 271–282.

6. J. Brandt, K. A. Quaid, S. E. Folstein, P. Garber, N. E. Maestri, M. Abbott, P. R. Slavney, M. L. Franz, L. Kasch, and H. H. Kazazian, "Presymptomatic Diagnosis of Delayed-Onset Disease with Linked DNA Markers: The Experience in Huntington's Disease," *JAMA* 261 (1989): 3108–3114; G. J. Meissen, R. H. Myers, C. A. Mastromauro, W. J. Koroshetz, K. W. Klinger, L. A. Farrar, P. A. Watkins, J. F. Gusella, E. D. Bird, and J. B. Martin, "Predictive Testing for Huntington's Disease with Use of a Linked DNA Marker," *New England Journal of Medicine* 318 (1988): 535–542.

7. K. A. Quaid, unpublished data from a survey of 23 testing centers in the United States, 1992.

8. Ibid.

9. R. Stern and R. Eldridge, "Attitudes of Patients and Their Relatives to Huntington's Disease," *American Journal Med. Genetics* 12 (1975): 217–223.

10. K. A. Quaid and M. Morris, "Reluctance to Undergo Predictive Testing: The Case of Huntington Disease," *American Journal Med. Genetics*, in press; K. A. Quaid, J. Brandt, and S. E. Folstein, "The Decision to Be Tested for Huntington's Disease," *JAMA* 257 (1987): 3362; D. Craufurd, L. Kerzin-Storrar, A. Dodge, and R. Harris, "Uptake of Presymptomatic Predictive Testing for Huntington's Disease," *Lancet* 1 (1989): 603–605.

11. M. Bloch, S. Adams, S. Wiggins, M. Huggins, and M. R. Hayden, "Predictive Testing for Huntington Disease in Canada: The Experience of Those Receiving an Increased Risk," *American Journal Med. Genetics* 42 (1992): 499–507.

12. M. Huggins, M. Bloch, S. Wiggins, S. Adam, O. Suchowersky, M. Trew, J. D. Klimeck, C. R. Greenberg, M. Eleff, L. P. Thompson, J. Knight, P. Mac-Leod, K. Girard, J. Theilmann, A. Hedrick, and M. R. Hayden, "Predictive Testing for Huntington Disease in Canada: Adverse Effects and Unexpected Results in Those Receiving a Decreased Risk," *American Journal Med. Genetics* 42 (1992): 508–515.

13. H. Brody, *Stories of Sickness* (New Haven: Yale University Press, 1987).

14. D. S. Karjala, "A Legal Research Agenda for the Human Genome Initiative," *Jurimetrics* 32 (Winter 1992): 121–222.

15. C. McKay, "The Effects of Uncertainty on the Physician-Patient Relationship in Predictive Genetic Testing," *Journal of Clinical Ethics* 2 (1991): 247–250.

16. Quotations come from voluntary responses to a notice placed in *The Marker* requesting at-risk individuals to share their thoughts about being at risk, about the test, and their reactions to the test results.

17. Brandt et al., "Presymptomatic Diagnosis," pp. 3108–3114.

18. P. R. Billings, M. A. Kohn, M. de Cuevas, J. Beckwith, J. S. Alper, and M. R. Natowicz, "Discrimination as a Consequence of Genetic Testing," *American Journal Human Genetics* 50 (1992): 476–482.

19. M. W. Shaw, "Testing for the Huntington Gene: A Right to Know, a Right Not to Know, or a Duty to Know," *American Journal Med. Genetics* 26 (1987): 243–246.

20. Huggins et al., "Predictive Testing," pp. 508–515.

The Human Genome Project and Human Identity

Dan W. Brock

Many of the ethical and philosophical issues raised by the Human Genome Project (HGP)—for example, maintaining the confidentiality and privacy of medical and genetic information or using information about a person's risk of acquiring particular diseases as a basis for exclusion from health insurance—are familiar in bioethics. The sheer amount of new information likely forthcoming from the HGP, however, may qualitatively transform these issues.[1] The three broad ethical issues that I will identify and briefly delineate in this essay are not part of the standard literature of either bioethics or law and medicine. Although these ethical issues are as yet not well defined in the literature on the HGP, they are no less important. Indeed, these issues implicate important legal principles and ethical practices with which we will have to grapple in coming years. Because the potential effects of the HGP that I will focus on are relatively long-term, my discussion will be somewhat speculative.

Each of these three ethical issues bears on our conceptions of human identity generally and on our own specific senses of identity as individuals. The first concerns issues of equality and, more specifically, equality of opportunity. The second concerns our conception of ourselves as responsible agents, a conception that underlies many of our moral beliefs and practices, as well as important legal practices in areas such as criminal law. The third ethical issue concerns the likely undermining of a clear standard of normality and its consequent effect on how we define ourselves and our identities in a psychological sense.

IMPLICATIONS FOR EQUALITY OF OPPORTUNITY

A commitment to equality has had a long and deep place in American politics and political philosophy regardless of how far we remain from realizing this commitment in practice. However, to think of equality as a single ideal is a mistake, because there are a number of competing alternative interpretations of how the commitment to equality should be understood. My focus here will be on equality of opportunity, which, in some interpretations, is widely accepted by those who disagree about stronger conceptions of equality.[2] Equality of opportunity is often contrasted with equality of result or outcome. For example, equality of opportunity with respect to a benefit or good is compatible with unequal results in the distribution of the good among those who had equal opportunity to acquire it. In this sense, equality of opportunity can be seen as a limited or minimal egalitarian ideal that is shared by those who disagree about equality in outcomes.

What We Mean by Equality of Opportunity

Equality of opportunity is open to several alternative interpretations. Equality of opportunity plays a role under circumstances of scarcity when not all individuals who want a particular good or position can have it. In one of its earlier and still important senses, the concept of equality of opportunity was used to attack legal or quasi-legal constraints on people's freedom to compete for scarce goods or positions. For example, legally enforced practices of racial segregation that excluded blacks from attending all-white schools and required them to attend schools of inferior quality, or hiring practices in which blacks or Jews were told they need not apply, violated this conception of formal equality of opportunity.[3]

While it may not be immediately apparent, formal equality of opportunity is compatible with preferential treatment of particular groups even though membership in those groups bears no relation to successful performance in the sought-after position. For example, a personnel office that considers all applicants for a job, but then gives preference to candidates who are members of a specific religion, does not violate formal equality of opportunity.

Fair equality of opportunity, however, adds two additional components to formal equality of opportunity. First, fair equality of opportunity demands that scarce offices and positions be filled on grounds that are reasonably related to successful performance in the particular position.[4] Thus, hiring tests or examinations are sometimes challenged as arbitrary on the ground that the knowledge or skills being tested do not differentiate applicants on the basis of future performance in the position.[5]

Even with the removal of all arbitrary grounds for selection, fair equality of opportunity is still not achieved. Fair equality of opportunity contains a second component—the removal of social and environmental barriers to success. For example, if the most desirable positions in a society required a college education available only at expensive private colleges that provided no scholarships or loans for poor students, poor applicants could correctly claim that they were being denied fair equality of opportunity. Though not formally excluded from college, poor students would lack an effective opportunity to obtain a college education, the qualifying condition for the most desirable positions in the society. Moreover, the means of gaining that education could be provided in the form of free or lower-cost public colleges or scholarships and loans for the more expensive private colleges. In these and other scenarios, fair equality of opportunity seems to require the alteration of *any* social or environmental factor that results in some segment of society's having a lesser chance of satisfying reasonable qualifying conditions for desirable scarce roles. When equality of opportunity is effected by ensuring the means to develop the necessary qualifying conditions for selection, people with roughly similar talents and abilities will have roughly equal life chances, opportunities, and expectations.

What Equality of Opportunity Could Mean

Even if social and environmental barriers and any other arbitrary grounds for selection are removed, fair equality of opportunity will still result in unequal outcomes between individuals. Outcomes will be unequal, in significant part, because of differences in natural abilities and the impact of these differences in the competitive world of equality of opportunity. Would the less-talented still have any moral complaint based on their lower life expectations? Suppose some of these less-talented people protested that, through no fault of their own, they were less able than most in some specific cognitive functions important to success in their society, despite having all available means for developing their cognitive capacities. These individuals would claim that this disadvantage denied them real or full equality of opportunity. It is important to note that if desired roles are truly scarce, there will *always* be qualifying conditions which have the necessary effect of excluding *some* persons from those roles. Would those persons who lose in this competitive process *always* have a complaint that their lack of the qualifying conditions denied them equality of opportunity? Could *any* difference between people that leads some individuals to lose competitions for scarce roles be protested as denying complete equality of opportunity to those individuals?

The convenient response to an individual's complaint that lesser cognitive ability led to a denial of full equality of opportunity is that there is simply no solution. More specifically, we respond by pointing out that nothing can be done about the residual differences between individuals that remain even after ensuring that every socially removable disadvantage has been eliminated. If an individual's cognitive limitation is due to his or her mother's poor nutrition during her pregnancy, which in turn resulted from her poverty, equality of opportunity would require that social welfare programs remedy the poverty and resultant malnutrition. However, if the cognitive limitation is simply a result of bad luck in the genetic lottery, nothing more can be done to achieve full equality of opportunity.

An important question at this point is whether equality of opportunity can be fully realized despite disadvantage in genetically based cognitive ability or whether, instead, equality of opportunity cannot be fully realized because of this residual cognitive disadvantage. We can see that the latter alternative is correct if we imagine isolating the genetic basis of the difference and then find that we could "correct" the disadvantage through some form of genetic therapy. If the therapy were so expensive that only the economically privileged could afford it for their children, equality of opportunity would require that the therapy be made available for disadvantaged children.

I hope the relationship of equality of opportunity to the HGP is now becoming clear. The HGP can be expected to increase dramatically our understanding of the genetic basis of a variety of properties that affect people's expectations in life. More specifically, the project may give us knowledge of the relationships between specific groups of genes and specific human properties and abilities. This knowledge may in turn lead to the development of gene therapy or engineering that could remove some of these genetic disadvantages, though we do not now know at what point in the future these new therapies will come. *No genetically based disadvantage will be immune from protest, and its removal sought, on grounds of equality of opportunity.* As Bernard Williams noted in a classic paper three decades ago, our notion of equality of opportunity collapses into equality understood as identity, because equality of opportunity requires the equalizing of *all* properties of people that affect their success in competitions for scarce goods.[6]

There will be two important results of equality of opportunity that will require deeper and more extensive equalization of people in the face of new genetic knowledge and subsequently developed gene therapies to correct deviations. The first result is that equality of opportunity will come into increasing conflict with other social and political values and institutions. It is already widely recognized that the full realization of the equality of opportunity that is now within our power is in deep conflict with the institution of the family.[7] Parents will try to secure advantages and opportunities for their children that will provide them with as good a life as possible. To eliminate the advantages and disadvantages that the family confers would, at a minimum, require deep intrusion into the family and would place severe limits on the freedom of parents to act on their natural love and concern for their children. More likely, such a goal would effectively eliminate the institution of the family as we now know it. The genetic knowledge and potential therapeutic capacities that the HGP will likely bring us will further strain our commitment to equality of opportunity and will force new decisions about the value of equality of opportunity relative to other social and political values and institutions.

The extension of equality of opportunity that is produced by new genetic knowledge and therapies will have a second and perhaps more profound result on human identity. In short, this extension will place pressure on individuals' psychological sense of identity. That sense is determined in part by a vaguely defined boundary around the self that distinguishes some properties of a person as internal and essential to the particular individual and, therefore, inviolate to manipulation, from other properties that are merely arbitrary and inessential external contingencies affecting the self. We may come to understand the partial genetic basis of many properties of individuals, such as abilities to do abstract reasoning like mathematics, or character traits like empathy for others' suffering, that are internal properties essential to our psychological conceptions of identity. However, to view these properties as merely an accident of the genetic lottery and therefore subject to manipulation by genetic therapy if they serve as barriers to equality of opportunity by disadvantaging us in our life expectations, is to see them as external and inessential to our identity. The resulting view is that of people as mere bearers of predicates, as Bernard Williams has put it, and distinct from all the properties that make people physically embodied, conscious beings.[8] I do not know how these tensions in our conceptions of individual identity will play out in light of the advances in knowledge and subsequent capacities for genetic therapy and engineering that the HGP will likely bring. However, I am confident that deep and difficult tensions in our conceptions of individual identity will arise.

IMPLICATIONS FOR OUR CONCEPTION OF HUMANS AS RESPONSIBLE AGENTS

The second aspect of human identity that the Human Genome Project is likely to affect deeply is our conception of ourselves as responsible agents and, more specifically, as morally and legally responsible for our actions, for the lives we live, and for the kinds of people that we become. The conception of ourselves as responsible agents is reflected in common moral beliefs and in important social and legal institutions and practices that place great value on individual self-determination.[9] By self-determination, I mean our interest in forming, revising over time, and pursuing our own conception of a good life. The exercise of self-determination involves making significant decisions about our lives according to our own values and aims. This conception of ourselves as responsible, self-reflective agents is embodied in our practice of holding ourselves morally and legally responsible for our actions. There is a long tradition of philosophical debate on the sense, if any, in which possession of free will is necessary to justify these practices of moral and legal responsibility.[10] Because analysis of that argument exceeds the scope of this essay, I shall only attempt to describe how the HGP is likely to affect and intensify the ongoing debate.

The Pervasiveness of Our Conception
of Humans as Responsible Agents

In criminal law. We need only consider the criminal law to illustrate how social and legal practices presuppose a conception of people as responsible agents. H. L. A. Hart has perhaps been the most prominent proponent of that view of the criminal law referred to as a "choosing system."[11] The law publicly announces that specified acts, such as homicide, assault, and rape, are not to be committed and holds people responsible for conforming their behavior to these requirements. Thus, the criminal law in effect commands people to conform their behavior within defined boundaries or suffer the punishment for their transgressions. For this command to be seen as an effective and fair means of controlling behavior, those to whom it is addressed must be capable of understanding the command and must have a reasonable opportunity to choose whether to conform to it. Only human beings, or more precisely persons, are believed to have the capacity required for the use of this means of social control. Dangerous animals must be controlled in other ways, such as by the preventive use of force, to render them no longer dangerous or harmful. The preventive detention employed with people who are dangerous by reason of mental illness also reflects assumptions about their lack of capacity for responsible behavior in this sense.[12]

In attitudes toward ourselves and others. A part of our complex practice of holding people morally responsible for their behavior and legally responsible for conforming to the criminal law is reflected in a related set of important attitudes about ourselves and others as responsible agents. The first-person attitudes we have toward ourselves and our own behavior include, most prominently, shame, guilt, and pride.[13] There is neither space nor need to analyze fully these complex attitudes here, but I will use the example of guilt to show how these attitudes presuppose that people who are justifiably the object of the attitude be responsible agents. Rational guilt entails the belief that an individual has acted wrongly and that it was reasonably within the individual's power not to have so acted. Experiencing guilt, or feeling guilty, is the way in which we acknowledge that we could and should have conformed to a legal or moral requirement when we have failed to do so. Guilt involves more than merely having certain feelings, as the common expression "feeling guilty" might suggest, and also more than having relevant beliefs. The experience of guilt involves complex dispositions to behave in a variety of ways, such as the dispositions to acknowledge our wrongdoing, to seek the forgiveness of those we have wronged, and to resolve to act differently on similar occasions in the future.

The practices of moral and legal responsibility toward others involve such attitudes or feelings as praise, blame, indignation, and resentment. These practices also exhibit complexities similar to those of the self-regarding responsibility attitudes. The practices of moral and legal responsibility, and their component feelings, attitudes, beliefs, and dispositions for behavior, are deeply embedded not only in institutions like the criminal law, but also in ordinary human life and social interaction. At the very least, however, they are in apparent conflict with a different way of understanding and responding to others' and our own behavior.

The Tension Caused by a Scientific View of Human Behavior

I will call the alternative view of people and their actions the scientific view of human behavior. Unlike the view of people as responsible agents, by which individuals' behavior is morally evaluated, the scientific view regards human behavior as a natural phenomenon that should be scientifically understood and explained.[14] The scientific view seeks the causes of human behavior and is usually not satisfied with superficial explanations of those causes. Instead, the scientific view seeks as deep and fundamental an explanation as is possible. For example, in physics this view takes the form of a search for and understanding of the most basic constituents of matter. In the science of human behavior during the age of modern genetics, this view will take the form of a search for the genetic bases of human behavior and of character traits and dispositions to behave in particular ways. While there may be special concern in the HGP for the genetic bases of human disease and with pathological behavior, there is no reason to believe that "normal" behavior is less genetically based than pathological behavior.

The scientific view of human behavior, of course, is not committed to the view that all human behavior is fully explained by its genetic origins alone. There is no reason to hold that view. Human behavior is explained by the interaction over time of an individual's specific genetic inheritance with a specific environment that will explain that person's behavior. Nevertheless, I believe that there is good reason to expect that the HGP will eventually enable us to understand human motivational and character traits as having important genetic determinants.

Will the Human Genome Project Alter the Tension?

What is the connection of this potential knowledge to the view of people as responsible agents and to the various practices that embody that view? It is only because the responsible agent's action is considered reasonably within the individual's control that we can justify holding that agent morally or legally responsible for conforming his or her behavior to moral or legal requirements. When we ascribe the action to the agent's character, it is nevertheless to ascribe the action directly to the individual. The newly discovered genetic knowledge is likely to be seen by many people as a threat to this view. Setting aside the possibility of changing an individual's genetic inheritance, one's specific and unique genetic structure is a paradigm of that which is viewed as beyond one's control and for which one cannot be held responsible. But that genetic structure will increasingly be viewed as an important determinant in explaining the behavior for which we will want to continue to hold the individual responsible.

If a person's genetic structure is a principal cause of behavior and that genetic structure is completely beyond the individual's control, can an individual justifiably be held responsible for the resultant behavior? To reward so-called good behavior will seem unfairly to favor individuals who are genetically predisposed to that behavior. Conversely, to punish "bad" behavior will seem unfairly to burden, or even to stigmatize, individuals who are genetically predisposed to that behavior. Moreover, the fact that behavior is shaped or influenced by an individual's environment or by the interaction between genetic structure and environment will do little to justify holding the person responsible for the behavior, since an individual's environment is largely beyond the person's control. The problem is apparent in the old adage, "To understand all is to forgive all." Once we understand the full and deepest causes of human behavior, we see that those causes are largely beyond an individual's control. This understanding will in turn throw into question the responsibility framework for viewing human action and behavior. After all, is it fair to hold people responsible for and punish them for behavior once we understand that it has a substantial genetic basis?

I noted above that the view of people as responsible agents is embodied in a wide range of attitudes toward ourselves and others, as well as in important social and legal practices. This view could hardly be more important to our conceptions of our identity as human beings. Yet to the extent that it comes under pressure from the knowledge that the HGP may yield, those attitudes and practices will be seen as resting on untenable foundations. The philosopher Peter Strawson suggested that these attitudes and practices are so deep-seated as to be largely ineradicable despite the pressure on them from the scientific view of behavior.[15] Even if this is so, to come to believe that these attitudes and practices have lost their foundations—even if we are unable to give them up—will place us in an intellectually uncomfortable and unstable position, the full consequences of which are likely to be profound however difficult to predict they may now be.

I emphasize that this problem of viewing ourselves as responsible agents hardly originated in the HGP. As already noted, there is a tradition going back many centuries of philosophical discussion concerning the problem of free will and determinism and their implications for moral and legal responsibility.[16] Even before we began to understand the specific genetic causal antecedents of human behavior in the manner that the HGP may further, it was commonly assumed that there were causal antecedents—both properties of people and of their environment—for all human behavior. Thus, the problem of reconciling this causal determination with a conception of free will necessary to sensibly and fairly hold people morally and legally responsible for their behavior was well recognized in advance of specific knowledge of those causal antecedents. Coming to know the specific genetic basis of those causal antecedents does not change the theoretical problem, but it is likely to increase the general public awareness of the problem and to call more sharply into question our practices of moral and legal responsibility. A debate that was previously confined to philosophers and philosophy classrooms may reach a much larger public and may put pressure on social and legal practices. When we can point to the specific genetic determinants of a behavior, instead of just raising a general theoretical worry that the behavior must have some such determinants, the impact on cultural attitudes and public debate about our practices of moral and legal responsibility is likely to be significant.

Many participants in the philosophical debate over free will and determinism assert that causal determinism is not incompatible with our view of ourselves as responsible agents and with our practices of moral and legal responsibility.[17] In my view, this "compatibilist" position, as it is called, is probably largely correct. But the very fact that this debate continues reflects what even many compatibilists acknowledge—a lingering worry that their defense of our practices of moral and legal responsibility may not fully answer all of the incompatibilists' concerns.[18] Our fundamental conception of ourselves as persons and, more specifically, our belief in ourselves as beings with sufficient free will to be responsible agents will come into question. It is certainly premature, and perhaps impossible, to predict with any confidence the outcome of this public debate and self-questioning of our conceptions of ourselves as persons. Nevertheless, the development of such a debate and questioning is another likely consequence of the HGP.

IMPLICATIONS FOR OUR CONCEPTION OF NORMALITY

The third area of human identity that the HGP will implicate concerns the relation of our conception of normality to our conception of our identities as persons. Because I am less confident about what the general implications of the HGP will be in this respect, and because I am also uncertain what the specific impact of this issue will be on the law, my remarks will be brief. I believe that a rough standard of normal human function figures importantly into the specific psychological sense that each of us has of our own unique individual identity as a person. Our senses of identity, in turn, clearly influence how others perceive us. This happens in two ways that I shall mention here.

Changing Conceptions of What Constitutes Good Health

The first way that a conception of normal human function figures into an understanding of individual identity is through the sense of being healthy on the one hand or diseased or sick on the other. Disease can be understood very roughly as a condition that limits normal human function.[19] The adverse functional deviation may affect a particular organ or biological process, but generally the condition will result in a functional disability at a behavioral level in one's life pursuits. How people react to being labeled as sick or diseased varies considerably, but such labels can often have profound and far-reaching effects on an individual's conceptions of self. Generally, it is when we have noticed an adverse effect or change in our normal functional capacities that we contact health care professionals and begin the process which can result in our being labeled as sick or diseased. Thus, the understanding of ourselves as sick or diseased tends to correlate fairly closely with our own and others' sense of us as having suffered some impairment of normal function.

Advances in genetic knowledge and screening have already resulted in the ability to predict, long before the onset of diseases such as Huntington chorea, that a person is likely or certain to develop that disease. The HGP can be expected greatly to enlarge these predictive capacities. The result will be that people who feel healthy and who as yet suffer no functional impairment will increasingly be labeled as unhealthy or diseased. This occurs even in the absence of genetic markers for disease when people who feel healthy are discovered to have risk factors, such as moderate hypertension, for the future onset of disease. When utilizing genetic markers, however, it is largely inevitable that there will be a substantial time-lag between our new capacities to predict that a person will have or has an increased risk to develop a particular disease and the development of effective therapeutic modalities to prevent or treat that disease. Thus, there will often be significant periods of time in individuals' lives when they will come to think of themselves as unhealthy or diseased though they neither feel sick nor suffer from losses of function. Moreover, in many cases no fully effective therapy will exist for an individual's condition. For many people, this labeling will undermine their sense of themselves as healthy, well-functioning individuals and will have serious adverse effects both on their conceptions of themselves and on the quality of their lives.

Changing Conceptions of What Constitutes Normal Human Functioning and Performance

There is a second, quite different manner by which our senses of identity are affected by the general social conception of normal human functioning. Our senses of identity are largely determined by the plans, aims, projects, commitments, and values that define our overall life plans. The important projects that we undertake and the commitments that we make are determined in part by our estimates of how well we can expect to fulfill them. This is so for at least two reasons. First, these projects commonly involve other people who require some adequate level of performance from us. For example, nearly all of us in the workforce require others either to hire and continue to employ us or to purchase our services if we are self-employed. Second, for most of us self-respect is significantly tied to the belief that we are and will continue to be reasonably successful in the pursuit of central plans and projects to the extent that this success is within our control.[20] Consequently, we usually do not choose to undertake projects and plans for which we know we are unsuited and which we will be unable to pursue with reasonable success.

However, "adequate" or "reasonably successful" performance is measured, even if only implicitly, against the performance of others. The competitive nature of the business world ensures this type of measurement in the work context. In other activities, such as sports, people may sometimes practice and seek to develop their skills largely on their own, but then they go on to "try out for the team" and to play against others. Even in less overtly competitive contexts, we implicitly, and probably in part unavoidably, measure our own accomplishments against the accomplishments of others. If we paint pictures or make furniture, we cannot help but judge our paintings and furniture against those that others have produced. The relation of our sense of our own significant abilities and capacities and our strengths and weaknesses to normal human functions is often very rough because numerous other factors can affect our senses of identity in many and complex ways. Thus, all I am claiming is that the relationship of our functional capabilities to normal human capabilities is only one significant factor that defines our sense of identity.

One important feature of the HGP will be its effects on this conception of normal human function. In some respects, the effect will be to sharpen and correct our present account. For example, new genetic knowledge about Alzheimer's disease shows us that much of the impairment of cognitive function that had been thought to be a part of the normal aging process is instead the result of a specific disease process. In other ways, however, the HGP may undermine our confidence about what normal human function is and may raise questions about whether it should continue to have its traditional importance in defining our senses of individual identity and self-respect.

If we gain the capacity to enhance human functioning in significant respects and on a widespread basis—for example, by genetic engineering or therapy that substantially enhances memory, or some physical capacity—the result may be to undermine our confidence in what is normal human function. Would these enhancements eventually redefine the norm of human functioning? Moreover, even if our account of normal human functioning does not substantially change, the HGP may raise doubts about whether we should define our senses of ourselves against that norm. If we develop important new abilities through genetic engineering or therapy to make widespread enhancements in people's functional capacities and capabilities, the unenhanced normal level may come to seem to us less significant as a standard against which we will measure ourselves and our own capacities and capabilities. It may in turn become a less secure basis on which to found our own self-respect. How secure will one's pride and self-respect be in a particular capability that, though it exceeds the unenhanced level of normal function, falls short of a widespread enhanced level of that capability produced by genetic engineering or therapy?

While the Human Genome Project holds out the prospects of exciting new knowledge of our genetic nature and of exciting new therapeutic possibilities, it is also likely to have profound effects on our conceptions of persons and on the individual sense of identity. To the extent that these conceptions of persons are deeply embedded in the law and in legal practices, we can also expect potentially profound changes there as well.

NOTES

This essay was previously published in substantially the same form in the *Houston Law Review* 29 (1992): 1–16. The copyright for the article is held by the Houston Law Review; the paper is reprinted with permission.

1. See Mark A. Rothstein, "Genetic Discrimination in Employment and the Americans with Disabilities Act," *Houston Law Review* 29 (1992): 23–84.

2. See Norman E. Bowie, ed., *Equal Opportunity* (Boulder, Colo.: Westview Press, 1988).

3. See Title VII, Civil Rights Act of 1964, 42 U.S.C. 2000e-e17 (1988); and Brown v. Board of Education, 347 U.S. 483, 493–495 (1954).

4. See, e.g., Washington v. Davis, 426 U.S. 229, 250–251 (1976); Greggs v. Duke Power Co., 401 U.S. 424, 436 (1971); and Gilbert v. City of Little Rock, 799 F.2d 1210, 1214–1215 (8th Cir. 1986).

5. Commentators have debated the legitimacy of such tests and examinations. See Mark Kelman, "Concepts of Discrimination in 'General Ability' Job Testing," *Howard Law Review* 104 (1991): 1158–1247; and James M. Conway, "Title VII and Competitive Testing," *Hofstra Law Review* 15 (1987): 299–322.

6. Bernard Williams, "The Idea of Equality," in Peter Laslett and W. G. Runciman, eds., *Philosophy, Politics and Society, Second Series: A Collection* (Oxford: Blackwell, 1962), pp. 110–131.

7. See James S. Fishkin, *The Limits of Obligation* (New Haven: Yale University Press, 1982), pp. 145–149.

8. Williams, "The Idea of Equality," pp. 128–129.

9. See Gerald Dworkin, *The Theory and Practice of Autonomy* (Cambridge: Cambridge University Press, 1988), pp. 12–20.

10. See, e.g., Gerald Dworkin's introduction to Gerald Dworkin, ed., *Determinism, Free Will and Moral Responsibility* (Englewood Cliffs, N.J.: Prentice-Hall, 1970), pp. 1–10; and Philippa Foot, "Free Will as Involving Determinism," in Bernard Berofsky, ed., *Free Will and Determinism* (New York: Harper and Row, 1966), pp. 95–108.

11. H. L. A. Hart, *Punishment and Responsibility* (New York: Oxford University Press, 1968), pp. 28, 44.

12. See Allen E. Buchanan and Dan W. Brock, *Deciding for Others: The Ethics of Surrogate Decision Making* (Cambridge: Cambridge University Press, 1989), pp. 311–317.

13. See John Rawls, *A Theory of Justice* (Cambridge, Mass.: Harvard University Press, 1971), pp. 453, 485–490; and Herbert Morris, ed., *Guilt and Shame* (Belmont, Calif.: Wadsworth Publishing Co., 1971).

14. See B. F. Skinner, *Beyond Freedom and Dignity* (New York: Knopf, 1971), pp. 3–25.

15. See P. F. Strawson, *Freedom and Resentment, and Other Essays* (London: Methuen, 1974).

16. Refer to n. 10 *supra* and accompanying text.

17. See, e.g., Daniel C. Dennett, *Elbow Room: The Varieties of Free Will Worth Wanting* (Cambridge, Mass.: MIT Press, 1984), pp. 101–102; and John Hospers, "What Means This Freedom?" in Berofsky, ed., *Free Will and Determinism*, pp. 26–27.

18. See Peter Van Inwagen, *An Essay on Free Will* (New York: Oxford University Press, 1983).

19. See Christopher Boorse, "Concepts of Health," in Donald VanDeVeer and Tom Regan, eds., *Health Care Ethics: An Introduction* (Philadelphia: Temple University Press, 1987), pp. 359–394.

20. See Rawls, *A Theory of Justice*, pp. 440–442.

Knowledge in Human Genetics:
Some Epistemological Questions

Michael Ruse

The beginnings of modern theories of heredity—"genetics"—are usually taken back to the 1860s, to the work of the Augustinian monk Gregor Mendel. Recent scholarship has, with some reason, questioned to what extent one can properly think of Mendel as a "Mendelian" and the degree to which he has been taken as a suitable icon for those of later times who needed a convenient historical predecessor. Be this as it may, it is true that Mendel's work was neglected until the beginning of this century, and—of pertinence to us here—true that Mendel himself seems to have had little or no interest in problems of inheritance in our own (human) species. He was interested in pea plants. In this narrow focus, he differed from others of his day—most particularly Charles Darwin—who may not have been thinking about heredity in a fruitful manner, but who were most definitely interested in the human question.

With the birth, or rebirth, of genetics in this century, the human question did start to become significant—very significant. As today's historians and sociologists of science have made very clear, it is important, in looking at the development of a new science or discipline, to ask certain basic questions about funding. It is all very well for a group of people to think that a certain area is ripe for development. Yet, any such system building will have a short life span unless its would-be practitioners can find ways to support their enterprise. There must be people—foundations, universities, governments—prepared to foot the payroll.

Now it seems clear that the new science of genetics was able, to a certain extent, to piggyback into existence on the backs of already-existing (and funded!) disciplines. For instance, T. H. Morgan at Columbia, who articulated the gene theory in 1911, came out of the vibrant American school of developmental embryology. Two areas in particular were receptive to genetics, agriculture and eugenics. Geneticists turned to these fields for support and consequently (not necessarily in any way illicitly) often shaped their products to meet contemporary interests and demands.

In agriculture, animal and plant breeders needed ways to improve their products, and they seized eagerly on the new science of genetics. They incorporated its ideas into their practices and supported further research in its methods and theories. Whether at the Busey Institute at Harvard or the John Innes Institute in England, that was where you found good geneticists. The wonder was not that Sewell Wright, a pioneer of population genetics, took his first job at the U.S. Department of Agriculture, but that not everyone else did.

Geneticists also found interest, encouragement, and support in eugenics. At the turn of the century, thoughtful people were worried—desperately worried—about the decline of civilization. They thought that biology was a key causal factor, especially through the outreproduction of the talented by the less gifted, and so they turned to genetics for confirmation, explanation, and solutions.

The human question, therefore, got tangled with genetics at an early point, and it has stayed thus tangled right up to the present. It is true that in the 1930s and thereafter, thanks especially to the rise of Hitler and his promulgation of vile and bogus racial doctrines, eugenics as traditionally conceived—especially eugenics done with an eye to the breeding of a superrace—has found little overt support. But under the guise of genetic counseling, many of the concerns of the eugenicists have lived on. Hitler's abuses did nothing to dent the intense interest in human genetics, as such.

Indeed, I would argue historically—and, what is important, conceptually (epistemically) also—that it is truly a mistake to think that there is genetics, and then there is human genetics. Whether from conviction or from the expediencies of funding and prestige, the two are linked together. Or, rather, the two are one, with human genetics being part of genetics per se.

A nice example of this symbiosis occurred in the English school of ecological genetics, led by the late E. B. Ford. Ostensibly, his interest was in Lepidoptera—butterflies and moths—and his school was famous (or notorious) for its naturalist attitude to science, as its members took happy holidays in the Scilly Isles, chasing after brightly colored insects with nets and killing bottles. But the money came from the Nuffield Foundation, and the rationale was that forces operating in the insect world—producing polymorphisms in wing color—supposedly throw light on forces operating in the human world—producing polymorphisms in human features. A triumph of the school, greeted no doubt with considerable relief by its members, was the confirmed prediction that the genetically controlled differences in human blood groups would prove to be associated with differences in susceptibility to diseases.

Turning the clock rapidly forward, it is even more the case today than it was in the past that the human side to genetics is large and important—drawing on the whole discipline for its very existence and progress, and in turn fertilizing and nourishing the subject as a whole. Molecular genetics is no less, perhaps even more, a subject with its human side. One thinks particularly today of the Human Genome Project (HGP), lurking over everything else, and yet still curiously dependent on the work on *E. coli*, not to mention Drosophila.

As a philosopher, no less than as a historian, when thinking of human genetics one thinks most immediately of value questions. I am no exception, and shall indeed mention value questions before I end this discussion. But first, and most prominently, I want to wrestle a little with epistemology—that is to say, with questions arising from the nature and theory of knowledge. What is the most pressing question of an epistemological nature arising from human genetics—remembering now that you cannot properly separate human genetics from the rest of genetics? Tradition and consensus seem to agree that it is "reduction"—so it is to this topic that I turn now.

ONTOLOGICAL REDUCTION

The term "reduction" bears striking similarities to the term "God"—there are as many meanings as there are people using it. Following recent discussion, I shall distinguish three main senses: ontological, methodological, and epistemological (this third is sometimes called "theory" reduction). I shall take the senses in turn, stressing that they have no definitive status and that there is bound to be inexact overlap between people's usages.

By ontological reduction is meant the thesis that there is nothing at the macrolevel which is not made up of parts existing at the micro- (or any specified) level. A baseball, for instance, is made up of pieces of leather and cotton and so forth, which are in turn made up of molecules, which are in turn made up of electrons and protons and other basic particles, and so on down as far as one can go. There is no more to the baseball than the parts of which it is made, ontologically speaking—that is, with respect to existence.

Genetics generally is clearly committed to ontological reductionism. The claim is not that every organism is made up of genes and nothing but genes. We know that that is false. Rather, the claim is that the overall physical body—animal, plant, or microorganism—is no more than the parts, of which the genes are a subset. Presumably, as a geneticist one believes that the genes carry the information for producing all of the parts, including themselves. But that already starts to take us beyond the strict claims of ontology. The point is that as living organisms we do not have bodily parts and *then* something else, whether you call it a vital spirit or élan vital or entelechy or even a "principle of organization," if you mean that this principle has some real being.

The human geneticist, as a geneticist, is committed to this assumption. Is there any reason why this should be cause for comment, or can we move at once to the next sense of reduction? Probably for most human geneticists there is little reason to pause for reflection—few today feel tempted toward any form of vitalism. Moreover, when you think of obvious candidates which violate ontological reductionism, their supporters generally hold back from pressing home the difficulties. I have in mind something like the immortal soul, which (the Christian presumably believes) stands in some state of true being above and beyond the molecules. My experience of believers these days is that usually they argue that any special ontological claims they want to make are of a sort that lie outside the realm of science, so there can be no conflict.

I suspect that the modern-day Cartesian mind/body dualist, like the philosopher Sir Karl Popper, rejects ontological dualism as I have categorized it. The dualist, of course, has lots of problems even without invoking modern genetics. However, genetics in general claims to speak to the whole organism—the moving, reacting, thinking organism—as well as to the pure morphology, and this applies to human genetics in particular. I do not think that anyone wants to say that the mind is simply made up of a number of parts—these parts having been produced by the genes—but the implication certainly is that the mind is an epiphenomenon of, or a different manifestation of, such parts.

The same holds also of something like free will. I will speak more to this topic in the next section, but the ontological implication of human genetics is that there is no such *thing* as free will. You do not have the body produced by the genes and then free will on top. This is not to say that humans are not free, but rather that the freedom must be sought within the bounds of ontological reductionism.

METHODOLOGICAL REDUCTIONISM

We come now to a far more controversial sense of reductionism. The methodological reductionist is one who, to borrow a famous phrase, thinks that "small is beautiful." Inspired by the history of science since the Renaissance, the methodological reductionist argues that the triumphs of science come through the revealing and understanding of ever-smaller entities of nature. Modern chemistry, for instance, would be quite impossible without an appreciation of molecules, and of the ways in which they can combine and react, causing new compounds and processes at the macrolevel of human experience.

Quite obviously, the (human) geneticist has a significant commitment to methodological reductionism. Although there is some question about whether Mendel himself actually believed in the existence of the "factors" he postulated as controlling his pea plants, the very rationale of twentieth-century genetics is to explain the larger—the phenotype—in terms of the smaller—the genotype. And the coming of molecular biology has only thrust this message home even further. Now we have the step down from the classical gene of Morgan to the macromolecules of Watson and Crick, and these macromolecules are themselves decomposed into smaller units.

I take it that the HGP is the apotheosis of methodological reductionism. In this century, human genetics—particularly on the medical side—has seen some triumphs of methodological reductionism. Sickle cell anemia, for instance, has been traced down to its etiology at the cellular level, and the genetic factors which produce it have been un-covered, even down to the ultimate molecules. Now, with the HGP, what we have is the attempt to reveal the whole of the human genetic story at the molecular level. The driving philosophy is that to know the person, you must know the parts.

Who but a spoilsport could object to any of this? And yet, one well-known geneticist (Francisco Ayala), by no means an opponent of re-search in human genetics, has written:

> The complete nucleotide sequence of the human genome might be helpful to biologists and health scientists as a data base for ex-periments. But I do not believe that it would contribute any more toward solving major biological or health problems than a computer printout of all the roads in the United States and of all the cars traveling over them in a particular year would help to ascertain the significant causes of highway accidents.[1]

Since the objections that are raised by Ayala and others are often di-rected against some real or apparent aspect of methodological reduc-tionism, let me now look at three potential problem areas, keeping hu-man genetics as the main focus.

First, there is the worry that the methodological reductionist (more precisely, the human geneticist committed to methodological reduc-tionism) will insist that *all* understanding comes from knowing the nature of the genes—or, in modern terms, the nature of the DNA. Implicit in this insistence is the belief that human beings—not just their physical natures but their behaviors and everything else—are simply the products of their genes. And with this belief, we are right down the road to some form of innatism or "genetic determinism," where everything is supposedly locked in by the genes—and thus one has a reason and excuse for all kinds of prejudice and discrimination, against women, blacks, and others.

I confess that when I hear enthusiastic discussions of the HGP, I start to share the worries of the critics. It seems like a too-easy path from claims about the genetic basis of (clearly horrible) diseases to claims about the inevitable inferiority of those with homosexuality-causing genes or alcoholism-causing genes or less-than-brilliance-causing genes, or whatever. However, even though one may get this impression from many discussions, what one can say is that if meth-odological reductionism is taken to imply that the genes are the sole causal factor in the finished human, then no human geneticist ought to be a methodological reductionist.

The geneticist's claim has always been that the adult phenotype is a function of the genotype in interaction with the environment. It is true that some gene effects are relatively stable across most encountered environments. It is also true that some environmental effects are relatively stable across most encountered genotypes. But, theoretically and experimentally, the end point of growth is a multicaused phenomenon. There is no such thing as a gene for alcoholism, if by this you mean that genetic theory insists that some people are determined by their genes to be heavy drinkers.

At the very least, then, as a good research strategy, methodological reductionism can hold only in a qualified way for (human) genetics. But, is it a good strategy even in this qualified way? There are those who, I suspect, would question even this. The feeling would be that, say what you like, with all of the good intentions in the world methodological reductionism in human genetics leads to innatism. I cannot reply to this charge here—indeed, as a philosopher I am not sure how I could ever reply to it, although I will touch on some of the concerns again at the end of my essay. However, I think one can (now defending the methodological reductionist) say something about an extreme form of the charge.

This is the critical claim that such reductionism, even if you allow some environmental component, nevertheless does point to some form of genetic determinism—where humans are illicitly being considered as automata, locked into what they do by preadult determining factors, of which genetic influences are a significant part. There is therefore no genuine freedom.

In response, let me say that there is no question but that a genetical approach presupposes some element of determinism. Even if we acknowledge the statistical nature of Mendelian thought, it still remains true that ultimately the geneticist believes that there are determinate causal laws. However, picking up on the discussion of the last section, while this kind of causal determinism probably precludes some ethereal entity of free will, there is a long philosophical tradition (compatibilism or soft determinism) which argues that genuine freedom is allowed by (even necessitates) this kind of determinism.

Without going into details here, let me simply point out that it is not required of the human geneticist that he or she suppose that we humans are behaviorally hard-wired like the ants. There are biological reasons to think we would have a degree of flexibility not possessed by the social insects, and there is no reason why our genes should prohibit (or work against) this. What I have in mind is a hypothesis that, for humans at least, the genes program us to think in certain ways—mathematical, logical, social, moral—but how exactly we act is a function of this programming and our specific problems or life situations. Thus I would argue that the methodological reductionism of human genetics does not necessarily lock one into a pernicious form of genetic determinism.

Let me move on now to the two other potential problem areas of methodological reductionism in human genetics. The first of these—well highlighted by the biologist Richard Lewontin and the philosopher Elliott Sober—is that such reductionism has a tendency to draw one's attention from the complexities and implications of genes in interaction. Sometimes it is complained that reductionists fail to take account of organization—virtually the defining characteristic of the living organism. At the stark level, this is simply not true. The molecular geneticist in particular is especially sensitive to organization, for the whole notion of the genetic code is fundamentally predicated on such organization.

But there is organization above the level of the gene in isolation that the reductionist stands in danger of ignoring. The Lewontin-Sober example is the already-mentioned case of sickle cell anemia. On the one hand, we have the emergent effect of an immunity to malaria which obtains only when we have a conjunction (heterozygote) of normal and sickle cell gene. A hard-line reductionist would be in danger of ignoring or overlooking this. On the other hand, we have the vital biological question of the reason(s) for the maintenance of such genes as the sickle cell gene in populations, given their deleterious effects (as homozygotes) on individuals. Only by looking at the group situation, in the macrocontext, where the sick homozygote is balanced by the super-fit heterozygote, can we start to answer such questions.

The conclusion to be drawn—at least the conclusion that I would draw and that I think Lewontin and Sober would generally draw—is not that methodological reductionism is necessarily a bad thing, but that as a research strategy it can answer questions only of a certain kind. And there are other important questions which need addressing.

The second of the other problem areas follows from this first. The methodological reductionist looks naturally at the individual—hoping to blow it apart and identify and understand its parts. This is what the HGP is all about. But this is necessarily to ignore or downplay the fact of human variation. And from there, it is but one step to belittling such variation. In the 1950s, there was an acrimonious debate between the geneticists H. J. Muller and Theodosius Dobzhansky over the status and significance of variation in populations—especially human populations—with the former arguing down such variation (the "classical" position) and the latter arguing it up (the "balance" position). Significantly, it was Muller who had made his mark by pushing a methodological reductionist strategy in genetics.

One can certainly argue that there is no reason logically why the reductionist should not be sensitive to group variation. The point is that the strategy does not prize such sensitivity and has a tendency to deaden it. If one thinks otherwise, then consider the HGP, which takes an ultraclassical position. It is true that the aim of the project is to aid in identifying variant "bad" genes, but these—as in any classical view—are set against the standard "good" genes. It is true also that there is usually grudging recognition of variation, but at once one is told that this involves 1 percent or less of the entire genome.

To which I can only respond on the part of the critic that it is precisely an assumption of methodological reductionism which leads one to think that a 1 percent phenomenon at one level translates into a 1 percent phenomenon at all levels. Human variation seems very important to us. Evolutionary biologists tell us that it is very important to us. Although I trust that in this section I have shown some major sympathy for methodological reductionism, I am not sure that one is justified on essentially a priori grounds in thus dismissing (or downgrading) the importance of such variation.

EPISTEMOLOGICAL REDUCTION

The final kind of reduction that I shall consider here—albeit somewhat briefly—is indeed that which has been the chief focus of philosophers. This is the "reduction" which is said to occur when one theory, or branch of theory, is taken into or absorbed by another theory, or branch of theory. Supposedly, this is what happened to Kepler's astronomical dynamics and Galileo's terrestrial mechanics when Newton published his gravitational theory. Such reduction is seen as being in opposition to replacement, as happened to phlogiston theory when the new chemistry arrived.

Traditionally, genetics was taken to be a paradigmatic case of such reduction—the theory of Mendelian genetics falling nicely beneath the theory of molecular genetics. But then, in line with a general questioning about whether such reduction ever does truly occur, critics swing the argument the other way. If reduction implies that the older theory can be shown to be a deductive consequence of the new theory, then neither in the Newtonian case nor in the genetics case did a reduction occur. Strictly speaking, the newer theory pointed up the crudities and inadequacies of the older theory.

Defenders of reduction, most notably Kenneth Schaffner, have argued back, agreeing that a deduction of old from new is not possible but maintaining nevertheless that there is something properly called reduction. The new theory, as in the genetics case, yields something much like the old theory, and—this is crucial—"corrects" the old theory, showing where and why it was wrong. The old theory is hardly discarded, but shown to be a step on the way to higher knowledge.

As you might imagine, the discussion has become somewhat technical—more so than is appropriate to introduce here. One point of relevance, however, is that even if reduction were possible, in genetics it will hardly be a neat deduction of one theory from another. Exceptions and variations are the keynote of biology. As Schaffner wryly remarks: "The picture of reduction that emerges from any detailed study of molecular biology as it is practiced is not an elegant one."[2]

For myself, let me simply say, agreeing with Schaffner, that for all of the difficulties, I am inclined to think that there are compelling historical and metaphysical reasons for persisting with the notion of reduction as deduction. If one does not, one is left with a sense of mystery as to why biologists (geneticists) clearly think there has been a continuity in their work as they moved from the old to the new.

However, and this will be a good point on which to end this section, even if one does think—and in a way, especially if one does think—that a traditional (although refined) reduction model applies to genetics, one should be wary of concluding that this gives carte blanche to all and any attempts to show how new understanding at one level translates into old understanding at another level. I think specifically in terms of human genetics, and again I have the HGP in mind.

First, that which was said just above about exception and variation applies no less, and in some respects more, to humankind than to other species. Hence, full understanding might be better obtained by going slowly, item by item, rather than by attempting one grand Baconian inductive leap over everything. What I am suggesting is that all that we now know about theory or epistemological reduction in genetics suggests that real insight comes when you work through one system (like the sickle cell case) rather than grab for everything, hoping that similarities will appear. They probably will not.

My second point is that no one is an island. Less metaphorically, no species exists just on its own, nor can insight come from the study of just one species. This brings us around full circle to what I said at the beginning about human genetics being part of genetics. Thoughts of epistemological reduction reinforce this point. If we have a theory reduction, then in some way one system of laws (or mechanisms) is being related to another system of laws (or mechanisms). But laws presuppose generality. One cannot stay with just one species. (There is some debate about whether this is a logical point. It is a true point, nevertheless.) Hence, an understanding of the genetics of humankind cannot be divorced from the understanding of the genetics of the rest of nature.

It is true that, since its inception, the HGP has become more sensitive to both of these points. It is now trying to produce more than just a road map of the genome. Its scope has been extended beyond the human. For once, I think, these are events which the philosopher would have predicted and prescribed. We did not have to be brought kicking and screaming to accept reality.

My conclusion is somewhat cautionary. My brief is to discuss epistemic issues pertaining to human biology. This I have tried to do, in what I hope was a fruitful manner. What I want to stress now is that you should not take my concern with epistemology as an endorsement of the thesis that this is what really counts when we consider human genetics—or even that we can in any true way separate the epistemological from other issues, notably the social-moral and the pragmatic. Clearly social-moral issues have been underlying just about all of my discussion, especially when I was considering methodological reduction. The very attempt to consider humans from a genetic perspective is obviously fraught with moral dilemmas.

Likewise pragmatic issues lurk. I am constantly amazed at the extent to which modern molecular biology is a technology-driven process. The results are interesting, but not earth-shattering. Getting the results is now where the true innovation lies. But this raises all sorts of pragmatic issues about costs and time and the like, including where best to concentrate one's inevitably limited resources.

I could go on about these matters, but where I want to conclude is simply with the realization that understanding comes at all levels and in all dimensions. The epistemological is merely one such level and dimension.

NOTES

1. F. J. Ayala, "Two Frontiers of Human Biology: What the Sequence Won't Tell Us," *Issues in Science and Technology* 111 (1987): 51–56.

2. K. F. Schaffner, "Philosophy of Biology," in M. H. Salmon et al., eds., *Introduction to the Philosophy of Science* (Englewood Cliffs, N.J.: Prentice-Hall, 1992), pp. 269–309.

Some Concerns about Self-Knowledge and the Human Genome Project

Panayot Butchvarov

Our topic is self-knowledge and the possibility that the Human Genome Project (HGP) will help us obey Socrates' injunction, "Know thyself!" What Socrates meant by this is not easy to tell, but I shall assume that the objects of the self-knowledge he called for are individual traits of character or dispositional psychological characteristics, especially the so-called virtues and vices, both moral and intellectual—for example, generosity and practical wisdom.

Now I want to sound a cautionary note at the very beginning. Biology alone cannot provide us with such knowledge. To know that a certain genetic structure is the basis of a certain particular dispositional psychological characteristic requires (1) the identification of the characteristic, which can only be done on the basis of behavior, verbal and nonverbal, (2) the availability of a classificatory, conceptual scheme that is sufficiently rich and subtle to match the expected richness and subtlety of the relevant part of the genetic map, and (3) the discovery of sufficient correlation between the genetic structure and the psychological characteristic to allow us to regard the former as the basis of the latter. None of these can be done by biology alone. The collaboration of psychology is required. And it is not at all clear that psychology is, or will be in the foreseeable future, advanced enough, with an adequately fine-grained conceptual framework, to be of sufficient help. Self-knowledge is quite a different goal from, say, discovering the genetic cause of Huntington disease.

But there is yet a second note of caution to be sounded. Most, if not all, of the psychological characteristics that would be the object of self-knowledge may have a genetic basis only in a very general and quite remote way, their specific nature being almost entirely due to environmental factors. If so, then knowledge of their genetic basis may contribute very little, if anything, to self-knowledge. To what extent this is true will again depend largely on what psychology will tell us.

So let us suppose that the HGP by itself has little relevance to our achieving self-knowledge. Nevertheless, its *psychological* relevance to our thinking about ourselves, its effects on our self-conception, can still be profound. In this respect an analogy with the theory of evolution may be instructive. It is debatable whether this theory has major philosophical or religious implications. Nevertheless, it has had an enormous effect on our conception of ourselves as a *species*. The research into the human genome can be expected to have at least as great an effect on our conception of ourselves as *individuals*, even if any inference from its results to strictly philosophical, religious, ethical, or legal conclusions (e.g., that hard determinism is true) should turn out to be fallacious. Indeed, in this respect another analogy, in fact almost a precedent, is available. Freudian theory can be interpreted, at least in its clinical applications, precisely as aiming at the patient's achieving knowledge of himself/herself as an individual. The fact that the theory is now held to have questionable merit as science is not the point of the analogy. The point is that Freudian theory has had an enormous impact on our conception of ourselves as individuals. Even uneducated people think of themselves as pushed by "unconscious drives," as victims of "complexes," for example. And many of those who have gone through prolonged psychoanalytic treatment regard the experience as profoundly revelatory about themselves. Self-knowledge of the sort the HGP is expected (however unwarrantedly) to make possible may have a far greater impact on the individual.

It is not difficult to illustrate what this impact could be. The *belief* that one has reliable and specific knowledge of one's intellectual capacities is bound to affect one's choice of a career. But, more important, it could affect profoundly one's self-esteem. We already know that persons' learning their IQ, or ACT, or GRE scores sometimes does have such an effect on them. But a belief that one has knowledge of one's own genetic makeup would have an effect enormously wider in scope and much more specific. The familiar examples I have just given may merely show that we are now unable even to imagine fully what will be possible.

From a philosophical standpoint, what seems especially interesting is the extent to which such a belief would encourage the adoption of hard determinism and, as a consequence, speculation on what one's life would then be like. It is a commonplace among philosophers that the free will–determinism issue has no practical implications, because any theory about it is compatible with whatever choices one in fact makes. But even if this is so, it does not follow that the *belief* that one has detailed knowledge of the causes of one's dispositions to make certain choices would also be irrelevant. It does not matter whether in fact one's choices have such a causal origin and whether, if they do, there is still room for free will. (One's genetic makeup would be at most causally necessary, but not sufficient, for any particular choices one makes.) What matters is whether a person possessing such a belief would almost inevitably think of himself/herself in a hard-deterministic way, and, more fundamentally, what it would be like to think of oneself in such a way. As Kant and many others have remarked, everyone who is engaged in thinking or choosing or making a decision presupposes that this thinking, choosing, or deciding are not causally determined in advance. We do not have a clear idea of what it would be for one to have *internalized* a fully deterministic conception of oneself.

The completion of the HGP could also have another consequence, which is chiefly ethical in nature. Unlike ancient ethics (for example, those of Plato and Aristotle), modern ethics (and political philosophy) has generally accepted the ideal of human equality with respect to treatment or at least consideration. And this ethical ideal has seemed compelling because of the empirical belief that there are no individual, ethnic, national, gender, or racial differences that would justify unequal, or discriminatory, treatment. We all are brothers and sisters, it is said. We all are children of God. We all are rational beings, inherently deserving to be treated as ends and never merely as means, Kant held. This belief could be strengthened (however unwarrantedly) by the results of the HGP, and if so our commitment to the ethical ideal in question would be strengthened. But suppose that the belief is not strengthened. Suppose that we are led to believe (again, however unwarrantedly) that there are highly significant (intellectual and moral) individual, ethnic, gender, or racial differences that are genetic in origin and that go far beyond the merely pathological. Would then the modern ethical ideal still be defensible? In what ways? And, whether or not it is, would it still be generally accepted? If it is not, what would be the social and political consequences?

Reductionism and Genetics:
A Response

William E. Carroll

Professor Ruse observes that modern genetics raises important epistemological and metaphysical questions about reductionism and determinism. Indeed, there is a tendency in some circles to think that genes are *the* determining elements or agents of human life, and in his discussion of what he calls "methodological reductionism," Ruse rejects such genetic determinism. I am interested, however, in the first kind of reductionism which he mentions, "ontological reductionism."

Professor Ruse notes that contemporary geneticists are committed to the view that a living body is nothing more than the sum of its material parts, and that genes are a subset of these constituent parts. He asks rhetorically, "Is there any reason why this [claim] should be cause for comment? . . . For most human geneticists there is little reason to pause for reflection—few today feel tempted toward any form of vitalism." I wish, however, to pause for reflection on ontological reductionism and its relationship to genetics. One important reason for pausing is Professor Ruse's observation that a commitment to ontological reductionism denies the existence of free will as some entity separate from the parts of a living human body; that is, free will cannot be something added to the body. If there is freedom for human beings, Ruse claims, it "must be sought within the bounds of ontological reductionism." The question of human freedom engages the attention of both Professors Ruse and Brock. In fact, the topics Professor Brock discusses depend upon a resolution to the questions Professor Ruse's essay raises.

Let me say at the outset that I do not think that free will is a separate entity, added to the body, but I do think that human beings have free will, and *yet* I think that it is impossible for human beings to have free will if ontological reductionism is true. It may well be the case, as Professor Ruse argues in his section on methodological reductionism, that genetics *itself* does not *necessarily* lead to *genetic* determinism, but I do think that once one is committed to ontological reductionism any notion of real human freedom is in fact excluded.

The crucial question is whether a geneticist or a biologist, or a chemist or a physicist, for that matter, *must* be committed to what Professor Ruse calls ontological reductionism. Are natural entities, including living beings, simply aggregates, the sum of distinct material parts? I think that one can study the chemical and biological structures of living things, and, in particular, genes and their relationship to the organisms in which they are found, without assuming the reductionist position. In fact, reductionism, especially ontological reductionism, is a claim about the nature of reality which is made in a more general study than that which comes under the purview of genetics, or biology, or chemistry. Contemporary biologists and chemists may not be particularly interested in the broader topics in such a general science of nature; they find the intricacies of their own specialties far more intriguing. In such specialties and subspecialties, including genetics, scientists have made extraordinary discoveries. But reductionism is not the claim of a specialist qua specialist: reductionism is not a claim, that is, of a geneticist as a geneticist. It is a claim about the nature of whole entities, precisely as whole entities. If we wish to investigate what the whole is—whether it is, for example, nothing more than the sum of its material parts—then we must move beyond the special studies of modern biology and chemistry. We must raise questions that are part of a more general science of nature.

When contemporary geneticists embrace reductionism—as many do—they share with their scientific colleagues a rejection of vitalism, or any other form of dualism, as Professor Ruse has noted. Vitalism and its philosophical parent, dualism, are claims made in the more general science of nature, to which I have already referred. Scientists see no evidence for some "mysterious life force," for immaterial substances, and thus they reject vitalism as unscientific. Since there seem to be only two fundamental possibilities, reductionism and vitalism, the conclusion follows that reductionism must be affirmed.

But are there only two possibilities? The history of science suggests otherwise. In fact, the history of science reveals that reductionist materialism and forms of dualism were evident in ancient Greece. Even what we now call "soft determinism" has its ancient counterparts in thinkers such as Epicurus and Lucretius. "Soft determinists" tend to identify human freedom with unpredictability, and they conclude that there is a compatibility between contemporary science's affirmation of the indeterminate or the unpredictable and human freedom. Although the existence of free will means that human actions are, in principle, unpredictable, it does not follow that actions are free *because* they are unpredictable. The "freedom" in "soft determinism" is really no different from the "atomic swerve" in Lucretius' *De rerum natura*. Human free will requires an immaterial faculty beyond the determination of physical change. Ontological reductionism, or any form of materialism, in principle excludes real human freedom.

In the face of ancient materialists and dualists Aristotle offered a third way. Aristotle would agree with the reductionists that a living substance is one material substance and not an aggregate of material and spiritual substances, as a dualist such as Plato would claim. But Aristotle would disagree with the reductionist's claim that the whole living organism is no more than its material components. For Aristotle, a living organism *as a whole* is a new actuality, an actuality not found in any of its parts. This new actuality is not the sum of the material parts. Nor is this actuality the structure of the parts. Aristotle calls this new actuality the substantial form of the living thing, or the soul, for the term "soul" means nothing more than substantial form, when the form in question is that of a living thing.

The soul for Aristotle is not something added to the material parts; he does not think that *first* there is a body and *then* there is a soul. Soul and body are not two separate *things*; they are two *principles* of one thing: the living entity. Aristotle's analysis of form and matter rejects both ontological reductionism and dualism. It provides a way to understand nature which affirms all the conclusions of modern science, including genetics, and allows us to account for free will as a function of the immaterial principle, the soul. Again, an appeal to an immaterial principle is not an appeal to the existence of an immaterial substance somehow joined to a body.

My point here is not to defend the adequacy of Aristotle's general science of nature—although I think that he offers the foundation of the only adequate account of nature[1]—but rather to show that there are options other than ontological reductionism and dualism, and that an investigation of these options is part of a general science of nature which is distinct from specialized sciences such as genetics. Such an analysis would show that there is no reason why a contemporary geneticist could not be an Aristotelian. In fact, I think that it would show that there are good reasons for any scientist's being an Aristotelian.

NOTE

1. Several contemporary authors have examined the relationship between Aristotelian natural science and modern science. Two of the most insightful are: Richard Connell, *Substance and Modern Science* (Houston: Center for Thomistic Studies, 1988); and William Wallace, *From a Realist Point of View* (Washington, D.C.: University Press of America, 1979).

The Human Genome Project and Epistemology

Evan Fales

The thought-provoking essays by Professors Brock and Ruse raise a good many thorny but important issues, spanning a range from philosophy of science to ethics, law, and policy. They also address epistemological issues, and I want to focus my comments initially on how *that* topic relates to some of the others. We can do this, I think, by asking the question, How much can we expect a perfect, or ideal, understanding of human genetics to tell us about human beings? And further, what reasons might we have for thinking that any particular answer to this question is true? (I am dispensing here with the question of whether or how we can come to have "perfect" knowledge of human genetics—not because that is not an interesting question, but because I assume that if we can obtain scientific knowledge at all, we can get it about genes, and because I also believe that if we can get any knowledge of genes, there is no reason in principle why we could not get all of it.)

I might begin by noting that the questions I have posed are intimately linked to the issue of reductionism which Ruse discusses. Let me say that I understand reductive explanations to be explanations of the characteristics of wholes in terms of the characteristics of their parts. There are various reasons why some philosophers would not be entirely happy with that way of construing reduction, but in the present context I think those concerns are not germane. Explanations of human characteristics in terms of the makeup of their genes clearly qualify as part/whole explanations; we can, if we wish, add for the sake of completeness the other physical parts of the system—including causal influences impinging from the environment. When those have been added in, what sort of explanations can we hope to have, and for what sorts of human characteristics can we hope to provide them?

Let me make a second aside before getting to this. Ruse distinguishes ontological, methodological, and epistemological reductive claims. Understanding reduction as explaining the nature of a whole in terms of the nature of its parts suggests that there are close connections among these three. The more successful we are at providing explanations of this type, the more credibility methodological reductionism acquires—though it may well be that not all scientific explanations are of this ilk. Successful part/whole explanations also lend credibility to ontological reduction: after all, if I can explain *all* the (nonrelational) characteristics of a whole in terms of the characteristics of its physical parts, what reason is there to think that the whole consists of anything *more* than those physical parts? Conversely, if a whole consists of nothing more than a structure of physical parts, it will seem a reasonable expectation that the nature of the whole can be wholly explained by appeal to its parts. Nevertheless, as I will argue in a moment, total explanatory reduction does not entail ontological reduction, nor is the converse true. Finally, reductive explanations in my sense just *are* what Ruse calls epistemological reductions, or at any rate, they are a subset of these. To give part/whole explanations for a class of objects is to be able to explain the nature of those objects, and the laws governing them, in terms of the natures of their parts and the part-governing laws.

Are human persons the sorts of beings all of whose characteristics are explainable in reductive terms? To ask this question amounts to raising the venerable mind/body problem. Or rather, it amounts almost to that. Pre-Cartesian Christians generally held, as near as I can tell—and people in many non-Christian cultures also hold—that the human person contains, in addition to a body and a mind, one or more further distinct entities called souls. Many *contemporary* Christians also think of the soul as some "third thing" distinct from the mind and the body, although other Christians clearly assimilate the soul to the mind. Unfortunately, I have not yet been able to make out what contemporary Christians in the former category *mean* by the soul. So I shall have to set that view to one side. I do, however, know my own mind (at least). So I'll stick to that.

Let's return, then, to the original questions, reformulated now as

(1) What sorts of human characteristics can be reductively explained (in terms of genes or whatever)? and (2) What sorts of reasons can be given in favor of any particular answer to (1)?

A considerable part of the promise of genetic explanations—I mean here the epistemic promise, of course, not the medical promise—arises from the often dramatic way in which certain identifiable phenotypic human characteristics (including mental disorders) have been linked to specific genes. But much credibility arises also from the *general* success and plausibility of reduction as an explanatory strategy throughout the sciences. In some quarters, anyone who suggests that there are limits to this strategy is looked upon, not without some reason, as an intellectual anachronism.

But there are, nevertheless, such limits—or, at the very least, there are conceptual puzzles which cast a deep shadow over the expectation that human mental characteristics can in general be given genetic (or genetic-cum-environmental) explanations of the usual reductive sort.

A fairly easy way to see one part of the puzzle is to recognize that no reductive explanation of the standard sort can be given of the computational capabilities of a computer. It does not matter how simple or complex a computing machine is: it can be one of the old mechanical cash registers that storekeepers used fifty years ago. Of course, we *can* reductively explain all the mechanical (or electronic, or whatever) behavior of the machine—all of the physical goings-on. But this by itself tells us nothing whatever about what arithmetical function, say, the calculator is calculating. It is only when we construe the physical input and output of the machine (and the physical operations that generate the one from the other) in terms of a system of symbols to which *we* assign semantic values (e.g., as referring to numbers) that anything can be said about the machine qua calculator. When we program a machine (by pushing number- or letter-labeled buttons, feeding it punched paper tape, or whatever), we assign meanings to these physical operations which allow us to interpret the ways in which we causally influence the machine as instructing it (say) to perform certain mathematical operations on certain numbers: similarly when the machine produces an output. But there is no *physical* feature intrinsic to the machine—or us—which determines that a given button codes for the number 2, or the operation of addition. Now it is notorious (1) that there is enormous latitude in the semantic values we can assign to input and output (or even, to what we take to *be* the machine's physical input and output) and hence one and the same machine can be said, under alternative construals, to be computing entirely different functions, or none; and (2) that machines built out of radically different materials and operating according to utterly different physical principles can, on suitable interpretations, be construed as computing the very same function. Thus, the functional characteristics of such a device are not reducible to physical ones.

Of course, one *can* reply that the machine does not really *have* any functional characteristics: these are only something that *we* impute to the machine because of the (arbitrary) semantic content we assign to its operations. However, we do, as I said, know our own minds. And there is not an arbitrary answer to the question of what *we* are thinking at a given moment. Yet we are, partly and after a fashion, beings who by conscious thinking perform computations. Insofar as our mental life includes reasoning and computations—and this is what artificial simulators of consciousness simulate best—it includes processes whose thought-content cannot be specified solely by specifying the physical structures and processes of the nervous system. So there is no reason to suppose that our mental life can, in any straightforward way, be explained by our genes and the genetically determined machinery of the brain.

The above line of argument is by now thoroughly familiar to philosophers. How much ice does it cut? Well, it does suggest that what correlations exist—if any—between brain mechanisms and thought, are contingent facts, and that these correlations must be established *independently* of our knowledge of brain mechanisms and the physical laws that govern them. But does irreducibility of the sort I have suggested entail ontological irreducibility? Does it entail mind/body dualism? And does it have any implications for the question of free will?

The line of argument I have all too briefly sketched does not, at least in any way I can discern, entail dualism. At least, it does not seem to require that the body and the mind (or whatever performs computations) are distinct substances. If it did, then either computers do not really compute anything or else they possess some kind of nonphysical thinking substance. What it might require, however, is that physical substances can, when properly structured, acquire properties that are not definable in terms of their standard physical properties.

When it comes to persons, I am strongly inclined toward a (property) dualism of sorts, though I have little sympathy for the view that persons consist of a physical stuff and a nonphysical stuff somehow pasted or patched together. However that may be, we can see now why (total) reducibility does not entail the falsity of substance dualism, and why, conversely, irreducibility does not entail substance dualism. The latter is at least strongly suggested by the argument just given. The former can be seen as follows: There is no reason why, for example, a person could not consist of a physical and a nonphysical substance, so joined that every characteristic of the nonphysical part is entirely parasitic upon the workings of the physical part. In that case, the nonphysical component plays no independent explanatory role, and reductive explanation would be possible. But substance dualism would still be true.

Even if all that I have said is true, it would be silly to suggest that knowing how a computer works physically can shed no light upon how it performs computations. If that were so, then a knowledge of physics would be of no help to computer designers. Similarly, it would be silly to suggest that knowledge of how the human body works can fail to illuminate how we think, why we develop certain personality traits, and so on. But how should all this knowledge affect our conception of ourselves as free agents, rational and morally responsible? How would a detailed knowledge of the underlying physical mechanisms impinge upon our image of ourselves as free choosers of courses of action? For of course there are "underlying mechanisms" in *some* sense. In part, the question is whether those physical mechanisms are connected via laws to conscious deliberation and the making of choices. I have argued that there are good reasons to deny this. Of course, the physics of the brain may constrain what we can effectively choose, but it is a familiar truism that physics constrains effective choice: I cannot effectively choose to jump over the moon.

More to the point, free will clearly requires *some* kind of determination of what is chosen. If the behavior-generating mechanism inside of me were a *random* behavior-generator—say one relying upon quantum effects—then I could not be said to control, or choose, my actions at all. Professor Brock worries that genetic determinism may require of us the conclusion, which Clarence Darrow already drew, that we do not have free will and cannot be held responsible for our actions. Ruse, on the other hand, suggests an antidote, and I think he is right. Just as an incapacity to reason or make choices can be (and sometimes is), unfortunately, genetically ordained, so too it is genes that ordain the sort of brain design that, in humans, is a necessary condition for the capacity to reason well and to freely choose.

Brock and Ruse seem agreed, however, that the only philosophical defense of human freedom left to us by our burgeoning scientific knowledge lies in the doctrine known as compatibilism. I think compatibilism is a dead end, although I cannot argue that here. Rather, I want to suggest a way to defend libertarianism—the view that some of our actions are genuinely free and (therefore) not causally determined. A libertarian can accept the view that a properly constructed brain is necessary for thought; libertarians *must* accept the view that thought—namely, the thinking involved in conscious acts of deliberating or choosing—sometimes determines a free act. If the only alternatives were between uncaused or random action and physically caused action, the libertarian view would have to be given up. But there is a *third* choice. Surely, those processes of deliberation that we call acts of making a choice are ones in which the early stages in our thought influence the succeeding ones. To think *randomly* would be to not deliberate or to choose at all. But why should we ever suppose that the relation of influence that exists between our successive thoughts in a train of deliberation is a relation of *physical* or *mechanical* causation? I can see no good grounds for thinking that at all, whatever our bodily mechanisms may be. Thus I think we can root for the success of the Human Genome Project—though on other grounds I am not sure we *should* do so—and still be good libertarians.

Commentary on Genetic Knowledge and Self-Understanding

Diana Fritz Cates

This symposium raises an important question in biomedical ethics: What is the relationship between self-understanding and the self's knowledge of its genetic makeup? To raise this question is, in effect, to raise four related questions: What *is* self-understanding? Why is self-understanding important? What might the self's knowledge of its genetic makeup, in particular, contribute to self-understanding? How important is this contribution? Let me pursue these questions briefly from the point of view of an ethicist.

What *is* self-understanding? Self-understanding is a kind of knowledge about the self. It includes a knowledge of the sort of person one is. It includes a knowledge of how one became the sort of person one is and how one might give deliberate shape to the sort of person one becomes in the future. It includes a knowledge of one's physical, personal, social, and historical conditionedness. It includes a knowledge of the possibilities that are open to one, given the peculiarities of one's conditionedness. It includes a knowledge of one's freedom to actualize realistic possibilities. It includes a knowledge of factors within the self that subvert the exercise of freedom.

Self-understanding is a kind of knowledge. Yet it is a knowledge, so to speak, not simply of the head, but also of the heart and the gut. That is to say, it is a knowledge that includes a certain feeling component. It is a knowledge of the sort of person one is and the sort of person one is becoming—but a knowledge that includes an affectional orientation toward this self in the process of becoming. It is a knowledge of the self's limitations and possibilities—but a knowledge that includes a suffering or an enjoying of these features as *one's own*. In other words, self-understanding is in part a coming to terms, as a whole person, with what one rightly believes to be the case about the self. This might be expressed by the suggestion that self-understanding includes an "evaluative" component.

To move beyond self-knowledge in a narrow sense and toward a broader sense of self-understanding requires asking and pursuing questions like these: What difference might information about my genetic construction make to the way in which I live my life? How might this information assist me in living a better life? How might it diminish my prospects? What is the point of my life, anyhow? Genetic information which is not processed in this fashion cannot promote self-understanding.

Why is this sort of self-understanding important? It is important because it is partly in its pursuit that human beings flourish. It is partly in the pursuit of self-understanding that human beings are able to realize their distinctively human potentialities in a well-integrated and satisfying way. To put it another way, self-understanding is important because it is a component part of the cultivation and exercise of virtue. To shape well our habits of perception, our characteristic ways of responding to what we see, our deepest desires for the good, our skills in deliberating about what we ought to do—to shape well these and other activities which are part and parcel of the exercise of virtue—we need to understand who we are, how we function, what we do well, what we do poorly, what we have control over, and what we must suffer as beyond our control. When we deliberately obscure things about ourselves that affect our exercise of virtue, we become less capable of shaping our moral agencies well.

Suppose we grant that self-understanding is important to the cultivation and exercise of virtue. How might the self's knowledge of its genetic makeup, in particular, contribute to self-understanding? It seems clear that knowledge of a particular genetic condition like Huntington disease *could* contribute to self-understanding in that it could inform an affected person about certain painful constraints that have been placed on the exercise of her moral agency. It could allow her to suffer the realization of her future deterioration in such a way that she is enabled to choose well what to make of the few good years ahead. It could allow her, in particular, to prepare herself and her loved ones ahead of time for the loss of her very capacity to prepare. It could help her and her family come to terms with the fact that she will, in the future, lose her capacity to come to terms with anything. This strikes me as an important contribution. Of course, as Kimberly Quaid has suggested, such a contribution would likely be made only if the genetic information were given within the context of genetic, psychological, and psychiatric counseling.

The decision to know or not to know about one's genetic makeup, at least in a case like that of Huntington disease, is an intensely personal decision. We should, in principle, honor a patient's well-deliberated decision not to know. Still, a counselor should help a person who is at risk come to see that, if she waits to be tested—if she waits until known symptoms begin to arise—it may be too late at that point for her to experience a coming to terms with herself, her future loss, and her death. True, she will eventually lose this self-understanding anyhow. But it would be good to have the opportunity to acquire it and to enjoy it for a time. It would be good to have the opportunity to bring a kind of closure to her life and to her relationships with loved ones before losing the capacity to engage in this sort of personal work. It would be good to have the opportunity to prepare herself, her friends, and her family for what is not simply an obscure possibility, but is instead a powerful and terrifying reality.

The good of helping the self and others come to terms with the impending loss of the self is, of course, only one good among others. How important is this good? How important is this kind of self-understanding in a case like Huntington disease? This is the crux of the issue. Clearly, coming to understand oneself as a bearer of the genetic material for Huntington disease *can* bring certain benefits to at least certain people who have the support necessary to reap these benefits. Still, how important is this self-understanding when weighed against the other goods at stake?

What other goods are at stake? Peace of mind? Could there be such a thing as peace of mind in the presence of continued doubt and suspicion? It would depend on whether one was able to quiet one's doubts sufficiently enough to enjoy one's life. From an ethical point of view, the attempt to quiet reasonable doubts about one's future is troublesome, especially when this attempt has potentially injurious effects on those with whom one shares a life. Undoubtedly, some measure of self-deception is necessary to sane living. We should be aware, however, that self-deception in one area of our lives easily develops into a characteristic pattern of responding to pain, a pattern that spreads quickly, roots itself deeply, and over which we subtly lose control. A pattern of avoiding pain in one area of one's life easily spills over into a pattern of failing to notice the pain in the lives of others.

Perhaps this spiral can be avoided by patients who choose deliberately and after much reflection not to be tested. After all, to engage in open, honest reflection on the many difficult issues at hand, to judge in light of one's particularity that knowledge of one's genetic status would, on balance, injure one's prospects at living well more than it would promote these prospects, and to choose therefore not to know *is* to reach a certain self-understanding. It is, in part, to know and to suffer one's limitations and the possibilities which are realistically open to one, given these limitations.

In the end, the patient must ask: Will knowledge of my genetic make-up promote my own self-understanding? Will this self-understanding help me to live well the years that lie ahead? How the patient answers these questions will depend on her understanding of what comprises a good human life. It will depend on what she values most in her life. It will depend on particular features of her moral agency and her social situation. There may be a presumption, morally speaking, in favor of promoting maximum self-understanding, but this presumption must be held in tension with a respect for privacy and the exercise of compassion.

Genes and Osteogenesis Imperfecta

Christine Carney

Let me begin by saying that I have been invited to contribute here because I am genetically unlucky. I was born with osteogenesis imperfecta (OI), which is a brittle-bone disorder. It is a defect in the Type I collagen cells that act as the glue that holds skin to muscle and makes bones hard. I recently did a paper in a class on the disease, and now I know quite a bit more about it. It is also known as the chalk-bone disease. I estimate that I have had about twenty fractures, the most serious being broken femurs and a broken pelvis.

One of the questions people ask me is how long have I known that I had the disease. I have always known; my parents never kept the information from me, because I had to be *careful*. If I had a dime for every time they told me that. . . .

My parents first noticed that something was wrong when I was ten days old; my leg broke as my mother was changing my diapers. I was hospitalized for a month. Then when I was two months old they took me to the Mayo Clinic in Rochester, Minnesota, and I was diagnosed with osteogenesis imperfecta tarda. The "tarda" meant that my condition was a milder one, comparatively. Some people who have OI sneeze and break their ribs. Luckily my condition is not as serious.

I have recently received genetic counseling, and I have learned that I have OI Type IV. Since my parents are of normal stature, they were informed that they do not carry OI. My condition was brought about by a new mutation. This news was a welcome relief for my sister, who is pregnant now. This information means that my brother and sister cannot carry OI, and it ended long family disputes about whose fault it was that I had OI. I always told my parents it was their fault they had a genetically mutant child. We joke a lot in our house. One has to have a sense of humor about it.

One question that was at the forefront of my mind before I had genetic counseling was whether I could have children. This made me very nervous. Recently a woman asked me if I would have children, knowing that I have a fifty-fifty chance of passing OI along to them. I said I sure would! I do not think that it has been that bad having this disease. There are many worse ones being discussed at this conference, like Alzheimer's and Huntington's. I just have to be very careful.

To my relief I found out that it is possible for me to have children. Of course, it would be a high-risk pregnancy, and I would have to have a special obstetrician. All of this I can handle.

Another question people ask is how easy it is for me to break a bone. I have fallen down many times and nothing has happened. Last March, however, I had a seizure and fell and broke my pelvis in five places. That was the most serious fracture I have had. I was in the hospital for a month. It took about a year for me to heal completely and be able to walk without a walker.

I use a mobilized cart to get around on the campus of the University of Northern Iowa, where I am a senior social-work major—English minor. But if I just have to go to the grocery store, then I walk without or with the walker, depending on how I feel.

Something that is interesting about my earlier years is that I had never met a person with a disability until I was a freshman in college. Then I met Michelle Holdorf, who showed me around the campus. Michelle was born with just one arm, and I figured that if she could go to college then I could too.

I have since gotten in contact with the OI chapter support group and have met many people of varying ages with OI. In fact there is a woman who has become a friend of mine who is my age and has OI.

Another question I was asked recently concerned bitterness I may feel toward my parents for having given me OI. My answer was no, I don't feel bitter about my condition because I don't feel that I have it that bad. I am glad to be here. I would feel sorry for anyone who would abort a fetus because it had OI. It is really not that bad. The doctor told my mom to "treat her like a china doll, but keep her normal." What a task! I think that they did a pretty good job.

II. The Possible Uses and Abuses of Genetic Knowledge

Is Human Genetics
Disguised Eugenics?

Diane B. Paul

As a historian of modern genetics, I am often asked whether human genetics represents disguised or incipient or possibly a new kind of eugenics. Those who pose the questions may not be certain how to define eugenics, but they are almost always convinced that it is a bad thing, one which should be prevented. Indeed, fear of a eugenics revival appears to be a principal anxiety aroused by the Human Genome Project (HGP), in Europe as well as the United States. Acknowledging this concern, project advocates insist that the mistakes of the past will not be repeated. Thus at the Human Genome I Conference at San Diego in October 1989, James B. Watson, the HGP's first director, told the audience: "We have to be aware of the really terrible past of eugenics, where incomplete knowledge was used in a very cavalier and rather awful way, both here in the United States and in Germany. We have to reassure people that their own DNA is private and that no one else can get at it."[1]

While almost everyone agrees that eugenics is objectionable, there is no consensus on what it actually *is*.[2] Indeed, one can be opposed to eugenics, and for almost anything. As Sir Isaiah Berlin remarked about the protean uses of "freedom," its meaning "is so porous that there is little interpretation that it seems able to resist."[3] To denounce eugenics is to signal that one is socially concerned, morally sensitive (and if a geneticist, perhaps worthy of public trust). But it does not predict one's stance on any particular reproductive issue.

In 1990, the International Huntington Association and the World Federation of Neurology adopted guidelines, based on the recommendations of a joint committee, for the use of predictive genetic tests.

The committee considered the refusal to test women who "do not give complete assurance that they will terminate a pregnancy where there is an increased risk" of Huntington disease to be acceptable policy.[4] Was the committee endorsing eugenics? Some would say yes, while most of its members would certainly be appalled by the suggestion.[5] But as we will see, there is no objective answer—nor can there be—to the question of whether such a policy constitutes eugenics.

At present, arguments about the relationship of eugenics and human genetics rarely converge. One person thinks they have nothing in common, while another considers the former merely an extension of the latter, not necessarily because they disagree on the facts (though of course they sometimes do), but because they employ the term in conflicting ways, often without noticing. After noting that prenatal diagnosis inevitably involves the systematic selection of fetuses, Abby Lippman charges: "Though the word 'eugenics' is scrupulously avoided in most biomedical reports about prenatal diagnosis, except where it is strongly disclaimed as a motive for intervention, this is disingenuous. Prenatal diagnosis presupposes that certain fetal conditions are intrinsically not bearable."[6] Conversely, most geneticists employ a narrow definition that identifies eugenics with a social *aim* and often coercive *means*. Both broad and narrow definitions serve a political purpose. The former associates genetic medicine with odious practices and thus arouses our suspicions; the latter dissociates it from these practices, and thus reassures.

Francis Galton, who coined the word "eugenics," defined it as "the study of the agencies under social control that may improve or impair the racial qualities of future generations, either physically or mentally."[7] However, it is less often identified as a science than as a social movement or policy, as in Bertrand Russell's definition: "the attempt to improve the biological character of a breed by deliberate methods adopted to that end."[8] After all, its practical applications, especially the sterilization laws adopted in America and Germany, and Nazi racial policies, explain why people worry about the relationship of eugenics to human genetics. Had its advocates confined themselves to study, eugenics would not now be a source of anxiety.

However, eugenics cannot be *defined* in terms of the social policies that account for its sordid reputation. We know from the historiography of the last decade that people in many countries and across a wide political and social spectrum advocated policies to genetically improve the "race." Eugenics has come to be associated with its most infamous practices in the United States and Germany, and hence with racism and political reaction. But there were eugenics movements of quite different character in France, Brazil, and Russia, among other countries.[9] Even in the Anglo-American world and in Wilhelmine and Weimar Germany, eugenics had a more diverse constituency and set of means and aims than one would imagine from the requisite reviews of the subject in discussions of "social implications of the new genetics." Some eugenics enthusiasts favored and others repudiated compulsion, while eugenics was invoked variously in support of capitalism and socialism, pacificism and militarism, patriarchy and women's liberation.[10] Its social content has been infinitely plastic.

Notwithstanding their disparate political perspectives, all eugenicists did agree that individual desires should be subordinated to a larger public purpose. Whether of the right, left, or center, they assumed that reproductive decisions have social consequences and thus are a matter of valid social concern. This history explains one conventional line demarcating eugenics from something else (often human genetics). Policies are characterized as eugenic if their intent is to further a social or public purpose, such as reducing costs or sparing future generations unnecessary suffering. Expansion of genetic services motivated by concern for the quality of the *population* would be eugenic by this definition, while the same practices motivated by the desire to increase the choices available to individuals would not be.

Unfortunately, this criterion requires a knowledge of motives, which may not be obvious and are sometimes mixed. Indeed, genetic services are generally justified on one of two very different grounds: that they increase the options available to families and/or that they reduce the burden of genetic disease in the community, thus saving money.[11] Dennis Karjala asserts that "all cost/benefit reasoning in the reproductive rights area is essentially eugenics."[12] That is a plausible claim, although it is not easy to determine the real intention(s) behind the expansion of prenatal testing and other genetic services. If a social purpose is the litmus test of eugenics, we must assess the importance of different aims, which are not always made explicit—and when they are, may disguise the truth. It is surely easier to defend abortion (which is the object of prenatal testing) in the language of choice than that of cost savings.

A definition in terms of *intention* is also at odds with the use of "eugenic" to describe the *effects* of individual action or social policy. The latter definition is implicit when a practice is characterized as eugenic because "we are effectively changing the gene frequency by lowering the number of offspring with 'defective' genes."[13] If consequences are properly described as eugenic, motive is no longer germane. Individuals do not ordinarily intend to benefit the gene pool by their reproductive choices. But private decisions may, taken collectively, have population effects. If the word "eugenic" appropriately describes consequences, and not just intentions, it casts a very wide net. It would make perfect sense, given this usage, to call abortion following prenatal screening "eugenics"—whatever the motivations for individual decisions or government funding. A definition broad enough to include unintended consequences will necessarily incorporate most medical genetics, or even individual mating decisions.[14]

The recent discussion of "back-door" eugenics implicitly depends on a definition of eugenics in terms of effects. A number of critics have warned of a resurgence of eugenics as the unintended result of individual choices. In their view, the real danger arises not from state policy, but from our increased ability to *select* the kind of children we want. Thus the new eugenics will result from a multitude of voluntary decisions, or even demands for tests and screens, rather than from social policy designed with eugenic aims in view.[15]

Most commentators, however, still restrict the term "eugenics" to policies pursued for a social purpose. They often add an additional criterion: there must be an element of coercion. According to these definitions, the state "interferes with," or "controls," or "imposes" particular reproductive options, as in one definition of eugenics "as any effort to interfere with individuals' procreative choices in order to attain a societal goal."[16] Eugenics is often demarcated from genetics by this criterion. Thus a participant at the 1991 International Congress of Human Genetics asserted: "Eugenics presumes the existence of significant social control over genetic and reproductive freedoms. Genetics does not require any special control over genetics or reproductive freedom."[17]

However, if we apply the label only to programs involving some form of coercion, we exclude a large number of individuals and policies ordinarily associated with eugenics. Many eugenicists, especially in Britain, stressed the voluntary nature of their proposals.[18] Virtually all "positive" eugenics (which seeks to increase the incidence of desirable traits, rather than reducing that of undesirable ones) would be excluded by this definition. Francis Galton would no longer be a eugenicist. Nor would H. J. Muller or William Shockley, notwithstanding their schemes for the (voluntary) insemination of women with the sperm of especially estimable men.

Moreover, there is no value-neutral answer to the question of whether a policy is coercive. Coercion has different meanings in different political traditions. To classical liberals (like John Stuart Mill or Isaiah Berlin) or libertarian conservatives (such as Milton Friedman), a decision is voluntary if there are no formal, legal barriers to choice. Freedom is thus defined negatively, as the absence of restraint. Coercion, on the other hand, "implies the deliberate interference of other human beings" with actions a person would otherwise take.[19] One is coerced only if actively prevented from attaining a goal. To liberals in the tradition of T. H. Green or John Dewey, however, as well as to socialists, coercion is not simply a matter of removing legal barriers: we are free to choose only when we have the practical ability to agree or refuse to do something. From their standpoint, a *situation*, such as economic need, may also be coercive.[20]

For conservatives, then, the potential parents of a severely disabled child are free to abort the fetus or bring it to term. From a contemporary liberal or socialist standpoint, choice may be lacking, given the medical and other costs of caring for such a child. In this view, parents could be coerced into aborting a fetus by the threatened loss of insurance coverage or lack of social services.[21] (This is not to imply that pressure would evaporate with national health insurance. Even in a socialized system, "if there is no confidence in the willingness of society to care for their child once they are unable to do so, parents may choose to terminate a pregnancy against their own wishes and beliefs.")[22] Whether parents are "free" to choose in these situations is a question that will necessarily be answered differently from different political standpoints.

Given competing (and sometimes implicit) definitions of eugenics, which themselves reflect larger (and unacknowledged) differences in political ideology, it is not surprising that arguments about whether an individual action, social policy, or unintended effect is or is not eugenics often fail to engage. The confusion would be reduced if definitions that do obvious violence to history (or contradict common sense) were excluded. A definition of "eugenicist" that bars Francis Galton is, on the face of it, absurd. It would also help if those who invoked the term at least specified what they meant by it. But the attempt to draw a line that clearly demarcates all policies acknowledged to be eugenic from those that are not will likely prove as fruitless as the analogous attempts to demarcate "science" from "non-science."[23] I would like to suggest another, potentially more productive tack. Let us ask what scenarios people actually fear when they express anxiety about a resurgence of eugenics, and to evaluate which of them (if any) are likely, which (if any) are possible, and which (if any) are improbable.

Concerns about eugenics typically fall in one of three distinct classes. The first is fear of direct government programs. Those alarmed at the prospect of state intervention often cite the Nazi experience, though they usually expect the analogue to be less brutal. Thus the biologist Salvador Luria questioned whether the HGP will transform "the Nazi program to eradicate Jewish or otherwise 'inferior' genes by mass murder . . . into a kinder, gentler program to 'perfect' human individuals by 'correcting' their genomes."[24] The activist Jeremy Rifkin, supported by some religious and disability rights groups, has called for a moratorium on human gene therapy research until such time as the NIH establishes a "Human Eugenics Advisory Committee" to evaluate its implications.[25] These critics fear a slippery slope leading to state action. They believe the government might try to design workers less susceptible to environmental insults, or to redesign us in other ways. "You could see genetic engineering of human beings from the fetal level on up," suggests Andrew Kimball, who directs Rifkin's Foundation on Emerging Technologies. "If they found that your child who's in kindergarten was predisposed to shyness, they would alter that child not to be shy. . . . These technologies have an enormously eugenic potential."[26]

How realistic is the fear of direct government intervention? In respect to coercive gene therapy, not very. Even if it were technically feasible, large-scale gene implantation would be an extraordinarily expensive kind of eugenics. And cost saving has always been the strongest motive for eugenics. In the early decades of the century, visitors at state fairs and expositions were warned of the price of leaving heredity to chance and apprised "that every fifteen seconds a hundred dollars of your money went for the care of persons with bad heredity."[27] College (and even high school) textbooks carried the message that "the cost of caring for those who cannot care for themselves because of their bad breeding is very heavy—perhaps two hundred million or more a year."[28] Economic considerations explain why only one of the thirty state sterilization statutes adopted between 1907 and 1931 extended to the noninstitutionalized—and why the rate of sterilization accelerated during the Depression.[29] Cost-benefit considerations today are less crude and, given sensitivities about abortion, sometimes less explicit. But they remain a principal incentive for state provision of genetic services.[30] And it will be a long time, if ever, before gene therapy or any form of "positive" eugenics can be promoted as a way to save money.

Cost-benefit analysis could, however, produce a stimulus for "negative" eugenics. After all, the plus side of the cost-benefit ledger is represented by the number of terminations achieved for a specified condition. The more women who are screened, and affected fetuses aborted, the more efficient the genetic service. Hence cost-benefit arguments for state support of these programs provide an incentive to expand genetic testing and "maximise the rate of terminations of pregnancy for 'costly' disorders."[31]

American states already sponsor many prenatal and newborn screening programs, including some that are mandatory. For example, every state tests for phenylketonuria (PKU); parental consent is rarely required.[32] Ultrasound screening is also routinely performed (without consent) on virtually every woman who sees a doctor early enough in her pregnancy.[33] Cost-benefit considerations help explain the vast expansion in the number of women who now undergo prenatal testing. They provide a powerful inducement to test more women, for more disorders, at an earlier age. As Neil Holtzman and Andrew Rothstein note, "Avoiding the conception of an infant at risk for a genetic disease—or avoiding the birth of a fetus prenatally diagnosed as having one—will often be less expensive than clinical management."[34]

Pressures to routinize screening will certainly increase with the development of tests that predict which individuals will bear offspring with genetic disorders. Individuals with the autosomal dominant gene for Huntington disease, polycystic kidney disease, and a rapidly growing list of other late-onset disorders can now be identified before they suffer symptoms and make reproductive decisions. Carriers of a number of autosomal recessive disorders, such as Tay-Sachs disease, sickle cell anemia, and most recently, cystic fibrosis (CF), can also be identified. CF is the most common severe genetic disease affecting Caucasians, with an incidence of about 1 in 2,500 live births. About 1 in 25 white persons (or more than eight million people) are asymptomatic carriers.[35] The ability to screen for common autosomal recessive diseases offers a solution to the problem that has bedeviled eugenicists since the first decade of this century.

Negative eugenics always faced a formidable practical barrier: the fact that most genes responsible for defects are recessive and thus hidden in apparently normal carriers. Policies such as sterilization and segregation only prevent the affected from breeding, and therefore work very slowly. Various schemes have been proposed to overcome this difficulty. In the 1930s, considerable attention was directed to identifying heritable antigens in the blood linked to genes for mental or moral defects, in the (disappointed) hope that these antigens could serve as genetic markers for the traits of interest.

Other schemes have relied on the phenomenon of partial dominance, that is, the fact that nearly all mutant genes in humans have some phenotypic effect. The most ambitious program of this type was proposed by H. J. Muller in 1949.[36] Unlike the early eugenicists, Muller understood that we all carry harmful genes. But he also realized we did so in different degree (to be precise, that the number of individuals carrying slightly harmful genes would form a Poisson series, with eight as its average). Partial dominance allows for the possibility of a potentially efficient means of selection. In principle, genotypes can be "surveyed" and those individuals falling in one tail of the distribution identified. Muller calculated that the genetic status quo could be maintained if the most mutant 3 percent of the population refrained from reproducing.

When Muller proposed this scheme, the technology of heterozygote detection did not yet exist. But in the same year James Neel untangled the genetics of sickle cell anemia and Linus Pauling the biochemistry, work that made possible the first sickle cell screening programs in 1971. Three years earlier, Pauling had suggested that all young people be tested for heterozygosity of the sickle cell and other deleterious genes and that a symbol be tattooed on the foreheads of those found to be carriers. "If this were done," he remarked, "two young people carrying the same seriously defective gene in single dose would recognize this situation at first sight, and would refrain from falling in love with one another. It is my opinion that legislation along this line, compulsory testing for defective genes before marriage, and some form of public or semi-public display of this possession, should be adopted."[37]

Sickle cell screening was strongly promoted in the African American community, where one in twelve persons is a carrier, and in a few states it was made mandatory. While the demand for these programs arose in part from within the black community, enthusiasm faded as the result of instances of discrimination against some of those identified as carriers. Screening programs for thalassemia and Tay-Sachs disease enjoyed much greater success (in part as the result of lessons learned from the problems with sickle cell screening). But these disorders are concentrated in particular ethnic groups—Americans of African, Mediterranean, and Ashkenazi Jewish descent. With the capacity to identify carriers of genes for common disorders, such as CF, the incentive for broad population screening has vastly increased. Wilfond and Fost note that "the potential for CF carrier screening programs will create an entrepreneurial opportunity that will dwarf all previous screening programs."[38] Indeed, biotechnology companies have begun offering tests, notwithstanding the 1990 and 1992 recommendations of the American Society of Human Genetics against routine screening where there is no family history of CF, and pilot programs to evaluate population screening are under way.

It is easy to predict that as we can screen more accurately and cheaply for more common disorders, the use of genetic tests will become increasingly routinized. Will individuals be pressured to make particular reproductive decisions as a result? That would qualify as eugenics by most definitions. It would in any case be a very unhappy development. And it is a likely one, though not as the result of *state* intervention. As Karjala notes, "Given the natural revulsion that most people feel for interference through mandatory testing or, even worse, mandatory abortion, the issues [of "genetic freedom and genetic responsibility"] are likely to be raised obliquely" through the health insurance system, HMO policies, or doctor pressure.[39]

A number of linked developments—the trend toward respect for patient rights in medicine, the rise of the women's and disability rights movements, the adoption of a broad jurisprudence of privacy and reproductive freedom—have converged to produce wide acceptance of the principle of reproductive autonomy.[40] Pauling's comments appear far more startling now than they would have in the 1960s. Of course, there are still individuals, of varying political perspectives, who express varying degrees of doubt about the consequences of allowing autonomy to trump other values.[41] An extreme standpoint is represented by Margery Shaw, past president of the American Society of

Human Genetics. Shaw replied no to the question of "whether or not a defective fetus should be allowed to be born" and expressed optimism that "parental rights to reproduce will diminish as parental responsibilities to unborn offspring increase."[42] But hers is a now unfashionable minority position. And even she would rely on tort liability—not legislation—to bring about the desired result.

In 1927, Justice Oliver Wendell Holmes commented, in his opinion upholding a Virginia sterilization order: "We have seen more than once that the public welfare may call upon the best citizens for their lives. It would be strange if it could not call upon those who already sap the strength of the State for these lesser sacrifices . . . in order to prevent our being swamped with incompetence. . . . The principle that sustains compulsory vaccination is broad enough to cover cutting the fallopian tubes."[43] Since the 1940s, however, the court has moved in a very different direction. In 1942, it unanimously overturned a sterilization law in an opinion that termed procreation "one of the basic civil rights of man."[44] The court further expanded the scope of privacy and reproductive freedom in *Griswold v. Connecticut* (1965), where it struck down a law prohibiting the use of contraceptives, in *Eisenstadt v. Baird* (1972), where it held that "if the right of privacy means anything, it is the right of the *individual*, married or single, to be free from unwarranted governmental intrusion in matters so fundamentally affecting a person as the decision whether to bear or beget a child," and of course in *Roe v. Wade* (1973). Even if *Roe* were reversed, it would be a long way back to Justice Holmes. Moreover, without access to abortion, there would be much less demand for prenatal diagnosis. State intervention would in any case be fiercely resisted both by feminists, who are committed to the principle of reproductive choice, and their opponents, who oppose abortion, and by the disability rights movement. It would have to defeat powerful, organized, and highly diverse social forces.

But as noted earlier, pressures can be indirect and involve actors other than the state. Holtzman and Rothstein remark that "although we may not soon reach the stage of compulsory eugenics legislation, denying health care coverage because of genotype could exert pressure on at-risk families to avoid having children with disabilities, despite the families' wishes."[45] In fact, these are the anxieties most commonly voiced when people say they fear "eugenics." They worry that tests, screens, and therapies will be introduced and promoted because they have the potential to generate profits for biotechnology companies, savings for employers and life and health insurers, and protection against malpractice suits for physicians—and that they will lack realistic alternatives to the decision to be tested or to abort a fetus identified as "defective" as a result of policies adopted by these quasi-public or private actors. The problems will intensify as carrier and predictive tests for common disorders (representing large markets) become more reliable. That a test is mandated by an insurer rather than the state does not necessarily make its consequences less drastic.

At present, few insurers make use of genetic tests. That situation will likely change as these tests become reliable and cheap and as more is learned about dispositions to common disorders, such as cancer and hypertension.[46] In 1989, the American Council of Life Insurance issued a report to its member companies (the largest providers of life and health insurance in the U.S.) to prepare them for the day when people routinely learn their genetic profiles. It suggested that when individuals have access to knowledge about predispositions to disease, the companies must also. Otherwise, individuals who know that they are at high risk of developing a disease will load up on insurance.[47] The courts are not likely to bar insurers from acquiring and using information on risks. Thus the U.S. Court of Appeals for the Fifth Circuit recently held that a self-insured employer could limit health benefits for AIDS after an employee was diagnosed with the disease, a decision that "could apply to people in whom genetic tests indicate a high probability of future disease."[48]

Decisions may also be strongly affected by new "standards of care" adopted by professional organizations in response to (exaggerated or real) fear of malpractice suits. In her survey of state legislation and current practice standards, Katherine Acuff notes that "legislation is by no means the only, or even the predominant, route to ensuring adoption of policies regarding prenatal or newborn screening. . . . [Many] testing procedures have been incorporated into routine medical practice based upon the pronouncements of professional organizations, without the spur of legislative mandate."[49]

Thus in 1985 the Department of Professional Liability of the American College of Obstetricians and Gynecologists (ACOG) alerted ACOG members to the purported need to inform *all* pregnant patients of the availability of maternal serum alpha-fetoprotein (MSAFP) screening. The physicians were told it was "imperative that every prenatal patient be advised of the advisability of this test and that your discussion about the test and the patient's decision with respect to the test be documented in the patient's chart."[50] As George Annas and Sherman Elias note, the rationale for the alert was legal, not medical: "to give the physician 'the best possible defense' in a medical malpractice suit premised on the birth of a baby with a neural tube defect."[51] Indeed, for a variety of reasons, including high false negative and false positive rates associated with the test and the need for appropriate counseling, ACOG had concluded that routine screening was of dubious value and should not be implemented in the absence of appropriate counseling and follow-up services.[52]

The doctrine of "informed refusal" (codified in the 1980 case of *Truman v. Thomas*) has also contributed to the expansion of testing. It is relatively easy to document informed consent. Moreover, there is essentially no (medical) risk attached to the procedure. If a woman is tested, there is thus no potential problem of liability for the physician. But it is relatively hard to document that she gave an informed "no."[53]

In the absence of public policy designed to prevent it, reproductive decisions will often be driven by the conjoined interests of powerful nonstate entities, such as physicians, lawyers, insurers, and biotechnology firms.[54] These are entities over which the public has limited control—precisely because they are private. In some formulations, this is a more subtle version of the concern about state action. But it may also be its converse. Thus Robert Wright suggests that the real threat is not a government program to breed better babies. "The more likely danger," he writes, "is roughly the opposite; it isn't that the government will get involved in reproductive choice, but that it won't. It is when left to the free market that the fruits of genome research are most assuredly rotten."[55] In any case, this package of problems is certainly real, whether or not it is labeled eugenics.

The third category is in effect the converse of the concern about indirect pressure (although they are not contradictory and indeed are sometimes linked). This is the concern often described as "back-door" eugenics. It reflects the fear that consumers themselves will demand genetic tests and treatments. Commenting on the potentially huge market for the CF carrier test, Wilfond and Fost note that "marketing could easily generate sufficient fear and anxiety so that many people would demand screening."[56]

Will women in fact demand perfect babies—or at least babies more perfect than their neighbors'?[57] This is where the controversy about sex predetermination enters. No one, as far as I know, thinks that the U.S. government will order or even encourage women to abort females, nor is there any reason to think that insurance companies, physicians, or counselors would find this outcome appealing. (If anything, the reverse is true.)[58] But the possibility exists that technical developments, including chorionic villus sampling (CVS), noninvasive procedures such as ultrasound and the recovery of fetal cells from maternal blood, and eventually preconceptual sex determination, will generate demands for access to sexing procedures. There is no evidence that American women in general are likely to use fetal sex information to disproportionately abort females. In fact, they have been quite resistant to abortion even for serious medical conditions (except where certain early death or severe mental retardation is involved).[59] The real problem, as Dorothy Wertz and John Fletcher have suggested, is that once the principle of choice in respect to a nonmedical condition is admitted, it is hard to know on what grounds it can be denied to parents who want to give their child a competitive advantage with respect to intelligence, height, or other socially desired characteristics.[60]

Whether we want to start down this road is a question worth asking, whether or not we subsume this choice under the term "eugenics." In my view, it is also the one we will have the hardest time thinking through, much less resolving politically. The reason, in brief, is this: those who are most concerned with these particular (mis)uses of genetics tend also to be the most committed to the principle of reproductive autonomy.[61] If the latter is considered an absolute right, we will have to accept a certain amount of "back-door" eugenics, under whatever rubric. A situation in which prospective parents can order the genetic characteristics of their offspring will also reinforce socioeconomic inequality, since those at the top of the scale can purchase (and for cultural as well as economic reasons want to purchase) more genetic services than do those at the bottom. This is not a future that many critics of the new genetic technologies would welcome. But it seems to be the path down which we are headed—unfortunately by default, rather than as the result of reasoned debate.

NOTES

1. Quoted in Joel Davis, *Mapping the Code* (New York: Wiley, 1991), p. 262.
2. More than twenty years ago, L. C. Dunn remarked in his presidential address to the American Society of Human Genetics that eugenics had tended "to become all things to all people." See his "Cross Currents in the History of

Human Genetics," *American Journal of Human Genetics* 14 (March 1962): 1–13; on p. 3.

3. Isaiah Berlin, "Two Concepts of Liberty," reprinted in his *Four Essays on Liberty* (New York: Oxford University Press, 1969), 118–172; on p. 121.

4. "Ethical Issues Policy Statement on Huntington's Disease Molecular Genetics Predictive Test," *Journal of Medical Genetics* 27 (1990): 34–38; on p. 37.

5. Thus Philip R. Reilly, one of the seven American members, is the author of *The Surgical Solution: A History of Involuntary Sterilization in the United States* (Baltimore: Johns Hopkins University Press, 1991).

6. Abby Lippman, "Prenatal Genetic Testing and Screening: Constructing Needs and Reinforcing Inequities," *American Journal of Law and Medicine* 17 (1991): 15–50; on pp. 24–25.

7. Francis Galton, *Inquiries into the Human Faculty* (London: Macmillan, 1883), p. 44.

8. Bertrand Russell, "Eugenics," in *Marriage and Morals* (London, 1924), 255–273; on p. 255.

9. For an overview of these lesser-known movements see Mark Adams, ed., *The Wellborn Science: Eugenics in Germany, France, Brazil, and Russia* (New York: Oxford University Press, 1990).

10. See Daniel J. Kevles, *In the Name of Eugenics: Genetics and the Uses of Human Heredity* (New York: Knopf, 1985) for an overview of Anglo-American eugenics.

11. For an example of explicit cost-benefit analysis, see J. C. Chapple et al., "The New Genetics: Will It Pay Its Way?" *Lancet* 1 (May 23, 1987): 1189–1192. See also Angus Clarke's critique of this approach, "Genetics, Ethics, and Audit," *Lancet* 335 (May 12, 1990): 1145–1147, and Sarah Bundey's letter in reply (June 9, 1990): 1406.

12. Dennis J. Karjala, "A Legal Research Agenda for the Human Genome Initiative," *Jurimetrics* 32 (Winter 1992): 121–222; on p. 160.

13. Ibid.

14. Elof A. Carlson, "Ramifications of Genetics," *Science* 232 (April 25, 1986), 531–532; on p. 531.

15. Troy Duster, *Backdoor to Eugenics* (New York: Routledge, 1990), p. x.

16. Neil Holtzman, *Proceed with Caution* (Baltimore: Johns Hopkins University Press, 1989), p. 223.

17. F. D. Ledley, "Differentiating Genetics and Eugenics on the Basis of Fairness," poster 1818, Eighth International Congress of Human Genetics, Washington, D.C., October 6–11, 1991.

18. On this point, see especially Richard A. Soloway, *Demography and Degeneration: Eugenics and the Declining Birthrate in Twentieth-Century Britain* (Chapel Hill: University of North Carolina Press, 1990).

19. Berlin, "Two Concepts of Liberty," p. 123.

20. Ruth R. Faden and Tom L. Beauchamp employ a strict definition in their chapter "Coercion, Manipulation, and Persuasion" in *A History and Theory of Informed Consent* (New York: Oxford University Press, 1986), pp. 337–381.

21. Benjamin S. Wilfond and Norman Fost, "The Cystic Fibrosis Gene: Medical and Social Implications for Heterozygote Detection," *Journal of the American Medical Association* 263 (May 23–30, 1990); 2777–2783; on p. 2781.

22. Clarke, "Genetics, Ethics, and Audit," p. 1146. He is writing of the situation in Great Britain.

23. The history of these attempts is traced in Larry Laudan, "The Demise of the Demarcation Problem," in Michael Ruse, ed., *But Is It Science?* (New York: Prometheus Books, 1988), pp. 337–350.

24. Salvador Luria, letter, *Science* 246 (October 13, 1989), p. 873.

25. Leslie Roberts, "Ethical Questions Haunt New Genetic Technologies," *Science* 243 (March 1, 1989): 1134–1136.

26. William Saletan, "Genes 'R Us," *New Republic* (July 17, 1989): 18–20; on p. 18.

27. Kevles, *In the Name of Eugenics*, p. 62.

28. W. E. Castle, J. M. Coulter, C. B. Davenport, E. M. East, and W. L. Tower, *Heredity and Eugenics* (Chicago: University of Chicago Press, 1912), p. 309.

29. The exception was North Carolina. See Reilly, *The Surgical Solution*.

30. B. Modell, "Cost/Benefit Considerations and Social Aspects of Genetic Services," Eighth International Congress of Human Genetics, Washington, D.C., October 6–11, 1991.

31. Clarke, "Genetics, Ethics, and Audit," p. 1146.

32. Katherine L. Acuff and Ruth R. Faden, "A History of Prenatal and Newborn Screening Programs: Lessons for the Future," in Ruth R. Faden, Gail Geller, and Madison Powers, eds., *AIDS, Women, and the Next Generation* (New York: Oxford University Press, 1991), pp. 59–93.

33. Lippman, "Prenatal Genetic Testing and Screening," pp. 21–22.

34. Neil A. Holtzman and Mark A. Rothstein, "Eugenics and Genetic Discrimination," *American Journal of Human Genetics* 50 (March 1992): 457–459; on p. 457.

35. Wilfond and Fost, "The Cystic Fibrosis Gene," p. 2777.

36. H. J. Muller, "Our Load of Mutations," *American Journal of Human Genetics* 2 (1950): 111–176. For a detailed discussion of this article see Diane B. Paul, "'Our Load of Mutations' Revisited," *Journal of the History of Biology* 20 (Fall 1987): 321–335.

37. Linus Pauling, "Reflections on the New Biology: Foreword," *UCLA Law*

Review 15 (1968): 267–272; on p. 269. See also his essay, "Our Hope for the Future," in M. Fishbein, ed., *Birth Defects* (Philadelphia: J. P. Lippincott, 1963), pp. 164–170.

38. Wilfond and Fost, "The Cystic Fibrosis Gene," p. 2777.

39. Karjala, "A Legal Research Agenda," p. 159.

40. For a discussion of the evidence for this claim see Diane B. Paul, "Eugenic Anxieties, Social Realities, and Political Choices," *Social Research* 59 (Fall 1992): 663–682; especially pp. 676–677.

41. Modest doubts are expressed by Alexander Capron in "Which Ills to Bear? Reevaluating the 'Threat' of Modern Genetics," *Emory Law Journal* 39 (Summer 1990): 665–696. Maura Ryan raises feminist objections in "The Argument for Unlimited Procreative Liberty: A Feminist Critique," *Hastings Center Report* (July-August 1990): 6–12.

42. Margery W. Shaw, "To Be or Not to Be? That Is the Question," *American Journal of Human Genetics* 36 (1984): 1–9; on pp. 1, 9. See also her essays, "The Potential Plaintiff: Preconception and Prenatal Torts," in A. Milunsky and G. Annas, eds., *Genetics and the Law II* (New York: Plenum Press, 1990), and "Conditional Prospective Rights of the Fetus," *Journal of Legal Medicine* 5 (1984): 63–116.

43. Quoted in Robert J. Cynkar, "*Buck v. Bell*: 'Felt Necessities' v. Fundamental Values," *Columbia Law Review* 81 (November 1981): 1418–1461; on p. 1419.

44. *Skinner v. Oklahoma*, 316 U.S. 535 (1942).

45. Holtzman and Rothstein, "Eugenics and Genetic Discrimination," p. 458.

46. Paul R. Billings et al., "Discrimination as a Consequence of Genetic Testing," *American Journal of Human Genetics* 50 (March 1992): 476–482.

47. Jerry E. Bishop and Michael Waldholz, *Genome* (New York: Simon and Schuster, 1990), pp. 297–298. For a useful discussion of this issue, see Henry T. Greely, "Health Insurance, Employment Discrimination, and the Genetics Revolution," in Daniel J. Kevles and Leroy Hood, eds., *The Code of Codes: Scientific and Social Issues in the Human Genome Project* (Cambridge, Mass.: Harvard University Press, 1992), pp. 264–280.

48. Holtzman and Rothstein, "Eugenics and Genetic Discrimination," p. 457.

49. Katherine L. Acuff, "Prenatal Newborn Screening: State Legislative Approaches and Current Practice Standards," in Faden, Geller, and Powers, *AIDS, Women, and the Next Generation*, p. 133.

50. American College of Obstetricians and Gynecologists, Technical Bulletin no. 67 (October 1982).

51. George J. Annas and Sherman Elias, "Maternal Serum AFP: Educating

Physicians and the Public," *American Journal of Public Health*, 75 (December 1985): 1374–1375; on p. 1375.

52. Ibid., p. 1374. Annas and Elias also note that only 40 percent of the women in one study reported discussing the test with a physician.

53. I am grateful to Karen Rothenberg for making this point.

54. One such policy would be explicit application of the Americans with Disabilities Act of 1990 to cases of genetic discrimination. But the ADA exempts insurers. Thus an employer may not "use the discriminatory practices of insurance companies as a pretext for refusing to hire, firing, or taking other adverse actions against an applicant or employee." However, insurers themselves may discriminate on the basis of demonstrable risk, if this practice is compatible with state law. Marvin R. Natowicz et al., "Genetic Discrimination and the Law," *American Journal of Human Genetics* 50 (March 1992): 465–474; on p. 471.

55. Robert Wright, "Achilles' Helix," *New Republic,* July 9 and 16, 1990, pp. 21–31; on p. 27.

56. Wilfond and Fost, "The Cystic Fibrosis Gene," p. 2777.

57. Robert Proctor argues that this is a spurious concern in "Genomics and Eugenics: How Fair Is the Comparison," in George J. Annas and Sherman Elias, eds., *Gene Mapping: Using Law and Ethics as Guides* (New York: Oxford University Press, 1992), pp. 57–93.

58. See B. Meredith Burke, "Counselors and Fetal Sex Selection," *Social Science and Medicine* (forthcoming).

59. See Ruth R. Faden et al., "Prenatal Screening and Pregnant Women's Attitudes toward the Abortion of Defective Fetuses," *American Journal of Public Health* 77 (February 1987): 1–3; Dorothy C. Wertz et al., "Attitudes toward Abortion among Parents of Children with Cystic Fibrosis," *American Journal of Public Health* 81 (August 1991): 992–996; and Jeffrey R. Botkin, "Carrier Screening for Cystic Fibrosis: A Pilot Study of the Attitudes of Pregnant Women," *American Journal of Public Health* 82 (May 1992): 723–725.

60. Dorothy C. Wertz and John C. Fletcher, "Fatal Knowledge? Prenatal Diagnosis and Sex Selection," *Hastings Center Report* (May–June 1989): 21–27.

61. Dorothy C. Wertz and John C. Fletcher, "Ethical Decision Making in Medical Genetics: Women as Patients and Practitioners in Eighteen Nations," in Kathryn S. Ratcliff et al., eds., *Healing Technology: Feminist Perspectives* (Ann Arbor: University of Michigan Press, 1990), pp. 221–241.

Molecular Biology: How Can We Translate the Laboratory?

Joseph D. McInerney

My assigned question—"The communication of molecular biology: How can we translate the laboratory?"—serves as an opportunity to reflect upon the general improvement of American science education. The vast majority of students in American schools will have their last formal exposure to science in tenth-grade biology, where we spend much time impressing upon them the details of science without leaving them with a sense of the importance of science for them personally or for society as a collective.[1] We certainly cannot do the latter without some of the former, but our tradition relies heavily on facts and details almost to the exclusion of their personal and societal implications. Even more distressing, our instruction rarely synthesizes the facts we present into a cohesive conceptual whole, and it largely ignores theory, which conditions our approach to research and to its applications. Perhaps this symposium will provide suggestions for making our science education more relevant for our students, while impressing upon them the need to understand major concepts in biology and, equally important, the nature and methods of science.

My recommendations about the translation of molecular biology necessarily reflect the perspective of one who develops curriculum for precollege biology courses, primarily for high school students. This may seem a relatively unimportant and inappropriate audience, given the potentially arcane analyses that the symposium topic, genes and human self-knowledge, can generate. One must remember, however, that today's high school students will be influenced in ways both obvious and subtle by the applications of our rapidly growing knowledge of molecular biology, and it is incumbent upon us to ensure that our students have the knowledge and skills to participate wisely as they help determine policies that will guide those applications.

To help ensure that the average student and the average adult possess that knowledge and those skills, we must be concerned with translating three aspects of the laboratory, not only for molecular biology, but for all sciences: (1) information, (2) technology, and (3) the nature of science. Last, we must consider the implications of our translation for teaching and learning.

INFORMATION

What do we want to say about molecular biology, given that we wish to avoid arcana that can find no ready contextual home in the mind of the average learner, whether teenager or adult? The title for this symposium provides excellent guidance for a conceptual focus because "human" and "self" direct us to two central, intimately related concepts in biology: evolution and variation. This is not a suggestion of convenience simply because the title is handy, but rather a reflection of the centrality of the concepts and of the opportunity to redress their unfortunate underrepresentation in the current biology curriculum.

A focus on evolution, of course, is essential to a cohesive curricular picture of biology.[2] In turn, an appreciation of variation within populations, what Ernst Mayr has called "population thinking,"[3] is central to understanding evolution and is one of the major intellectual contributions of the Darwinian revolution.

Equally important, a basic understanding of variation is essential to an understanding of human genetics,[4] of the biological context of much modern disease,[5] and of the attendant assumptions about policies that emerge to deal with such disease.[6] Molecular biology increasingly illuminates the range of genetic variation in the human population, and the Human Genome Project (HGP) promises to accelerate this interesting examination of biochemical individuality. A major data base at Johns Hopkins is devoted to organizing the information, and a new journal, *Human Mutation*, is dedicated to chronicling human genetic variation and exploring its biological significance. One has the sense that Archibald Garrod, who first noted the significance of human biochemical individuality some sixty years ago, would be gratified indeed.[7]

What do we want students to know about molecular biology and about its relationship to the concepts of human and self, and to evolution and variation? I shall begin with some very modest proposals for apprising students of the scope of molecular biology and follow those proposals with some suggestions about the concepts of human and self. The principles I propose are such basic working knowledge that you may protest at their simplicity and argue for a more elaborate understanding of the details of cutting-edge molecular biology. It is important, however, to remember John Dewey's caveat that facts, like fish, do not keep well. The education literature confirms that a concentration on basic principles, with attention to preexisting concepts in the mind of the learner, will have more enduring impact for the nonbiologist.[8]

What Is Molecular Biology About?

When molecular biology makes its way into the public consciousness through the media, DNA is most often the context, and I shall, therefore, focus on DNA as I consider how best to translate the molecular biology laboratory for the average person. Although molecular biology encompasses far more than the study of DNA, the public's exposure to DNA in the media, the symposium's focus on genes, and the limitations of space make a concentration on the genetic material itself prudent.

Principle 1. *DNA is an information molecule that is stable, replicates with great fidelity, and carries the information responsible for the full range of biological variation we see on the planet.* The information contained in human DNA, the result of at least three and one-half billion years of evolution, helps define what we are as a species and as individuals. It is responsible—in conjunction with complex cellular processes—for the regulation of virtually all organismic functions as well as the production of the organism's constituent parts, and is transmitted from generation to generation in the recurrent cycle of reproduction.

Principle 2. *New technology allows us to examine the genetic material in great detail and to manipulate it.* The advent of genetic technology in the early 1970s provided us with an unprecedented ability to study and manipulate the genetic material of all organisms, including humans. This ability allows us to attempt solutions to human problems (for example, genetically engineered bacteria, gene therapy, and DNA-based population screening) and to gain insights into enduring biological questions (for example, evolution, development, and regulation) that were impossible before the development of technologies such as restriction analysis and the polymerase chain reaction.

Principle 3. *Our increasing ability to analyze and manipulate genetic material raises profound questions of ethics, law, and public policy.* In genetics, as in all sciences, "science and technology can tell us what we *can* do, but not what we *should* do."[9] Decisions about the personal and social uses of molecular biology will require consideration of complex interactions among science, ethics, and public policy.[10] Students should understand the basic dimensions of those interactions so they can make informed personal decisions and, one hopes, participate effectively as they help determine public policy.[11]

Principles Related to "Human"

Principle 1. *Investigation of DNA from humans and many other organisms affirms that we owe our uniqueness as a species to the same processes of selection and descent with modification that produced all other species on the planet.* In short, all organisms share the same type of information molecule, and all life, including human life, "shares a single, complex history."[12] Information from molecular biology affirms and extends broad concepts of evolution, selection, and descent that biologists first determined through descriptive natural history and studies of morphology.

Although we share a biological history with all other species, the special influence of cultural evolution renders us unique among species and has provided us with unprecedented power to influence the course of both biological evolution and the physical evolution of the planet.[13]

Principle 2. *Molecular biology and evolution help us to understand both the commonalities and differences that constitute the "human."* Selection has helped shape the characteristics of the human population and, as for all other species, has set the limits of workable genetic variation. Most humans occupy the middle ranges of distributions for common characteristics because those ranges likely were more adaptive than were the ranges that represent outliers on the distributions.[14] The genes that contribute to middle-range constitutions therefore are more common in the population than are genes that code for the outlier constitutions. Because extreme divergence from "normal" probably has been disadaptive, extreme outliers "stand out by virtue of their appearance and their rarity."[15]

Geneticists, of course, acknowledge a great deal more normal variation than does the public (and especially the teenage public), which generally equates "normal" with "average." This simplistic calculus, which is entrenched in public perception, represents an educational challenge for all of us in science and science education.

Principle 3. Molecular biology has helped demonstrate that the current genetic constitution of modern Homo sapiens *reflects a paleolithic heritage, which we now subject to a markedly different range of selective pressures from those that existed thirty-five thousand to twenty thousand years ago.* The incongruence between our genetic heritage and our current environment likely is responsible for much of the morbidity and mortality we experience, especially in developed countries.[16] As Burkitt and Eaton assert, "The late Paleolithic Age, from 35,000 to 20,000 B.C., may be considered the last time period during which human physiology and biochemistry interacted with extrinsic influences typical of those for which they were originally selected."[17]

Principles Related to "Self"

Principle 1. Just as molecular biology helps to explain our uniqueness as a species and our relationship to all other organisms, so too molecular biology can help explain the uniqueness of each individual. While each of us is human, no two of us are the same. Indeed, as Thomas Caskey has shown (see figure 1), the range of potential human genetic variation is so staggering as to be almost incomprehensible.[18] Individual phenotypic variation is enhanced by the interaction of one's genetic individuality with a host of environmental variables.

Principle 2. Research in molecular biology continues to demonstrate the importance of human variation in the onset of chronic disease. The literature confirms the contribution of genes and environment to chronic, multifactorial disorders such as heart disease, cancer, diabe-

FIGURE I. THE EXTENT OF HUMAN GENETIC VARIATION

> 1. Any two people differ in about 0.1 percent of their DNA bases. Thus, 0.001 (0.1 percent) $\times 3 \times 10^9$ base pairs $= 3 \times 10^6$ base pairs of variation between any two individuals.

> 2. There are approximately 5 billion (5×10^9) people on earth. Thus, there are approximately $3 \times 10^6 \times 5 \times 10^9$ $= 15 \times 10^{15}$ base pairs of potential variation in the human population.

tes, and affective disorders.[19] For most such maladies, however, it is not yet clear which genes or combinations of genes are responsible, or how those genes behave in various environments.[20] Some biologists and physicians have suggested that a focus on Darwinian thinking and individual variation—rather than on individual diseases—will come to be the prevailing medical model,[21] but there is little evidence that such thinking is as yet pervasive in medical practice.

Principle 3. *Recognition of the range of normal variation in the human population affords us the hope of viewing human differences without stereotype.* An increased appreciation of "self"—of uniqueness—that derives from molecular biology should help students understand that variation is to be valued and that racial and ethnic stereotypes are without basis in biology. There is, for example, often more variation *within* members of a given racial group than there is *between* two different racial groups. Similarly, those who have developmental disabilities are not qualitatively different beings from those who are not disabled; rather, their disabilities represent a different point on the continuous distribution of human variation.

TECHNOLOGY

Most people never encounter science as scientists know it and discuss it; they encounter technology. Rarely, however, does the average person encounter technology related to molecular biology. He or she may encounter technologies *intermediate* to the application of molecular biological techniques—amniocentesis or chorionic villus sampling, for example—and those technologies are mystifying enough. Once the samples extracted by such technologies leave the clinic, however, the laboratory techniques to which they are submitted are virtually unknown to the layperson.

I propose that it is unimportant for the layperson to know about those technologies in detail. It is important, however, that the public understand some basic principles about the nature of technology and about the ways in which those principles relate to molecular biology and to the knowledge generated by its attendant technologies. The public should understand such concepts if only to be armed with appropriate skepticism about the capabilities of the technology itself and of the soundness of the resultant interpretations.

The American Association for the Advancement of Science, through the publication *Science for All Americans*, provides helpful guidance about the principles of technology that are important for general scientific literacy.[22] These principles certainly are relevant to the translation of the molecular biology laboratory.

Principle 1. *Technology extends our senses.* Many technologies associated with science help us see, hear, or touch objects or phenomena that we otherwise would be unable to experience. We must, however, understand the limitations of our technologies and the role of inference in the interpretations of information we derive from them. For example, much of our knowledge of deep space is based on inferences we derive from visible light and other types of radiation that reach the earth and that are collected by various instruments; we can neither see nor feel pulsars, quasars, or black holes. Similarly, we infer the presence of mutations in the genes for sickle cell disease or cystic fibrosis on the basis of restriction fragments displayed on a Southern blot; we can neither see nor feel the mutations themselves.

Principle 2. *Technologies often have unintended consequences.* All technologies (save those that arise by accident) are developed for a specific purpose, yet many have side effects that are unintended and, worse, unanticipated. Television brings news, entertainment, and knowledge of distant places to millions, but it also inures us to violence and may reduce academic achievement among our children. DNA analysis provides insights into longstanding biological questions such as gene regulation and evolution and allows us to detect genes associated with disease, but it has raised the specter of genetic discrimination to levels heretofore thought unimaginable.

Principle 3. *All technologies are fallible.* The examples are legion. With respect to molecular biology, the public must know that laboratory techniques can fail for reasons that range from those inherent in the technologies themselves to those associated with human error. When technology fails in the context of basic research in molecular biology, the results are disappointing and frustrating for the investigator. When technology fails in the context of molecular-genetic medicine, the results can be catastrophic for individuals and society.

Principle 4. *Technology and science are interrelated and interdependent, but they also are different.* In many disciplines, including molecular biology, science and technology are so interrelated that it is difficult to know where one begins and the other ends. It is helpful, however, to distinguish the two for the public, because they have inherently different objectives. At Biological Sciences Curriculum Study (BSCS), we distinguish them as follows: "Science proposes explanations for observations of natural phenomena; technology proposes solutions to problems of human adaptation to the environment."[23] The choice of the verb "proposes" in each instance is quite deliberate; it helps to emphasize the tentative nature of scientific explanations and technological solutions.

NATURE OF SCIENCE

This category of public understanding, while perhaps the most difficult to develop, is without question the most important. Nowhere is American scientific illiteracy more evident, and nowhere more debilitating and dangerous, than in the failure to distinguish "science as a way of knowing" from other ways of knowing about the world.[24] The public's inability—or unwillingness—to distinguish the nature and methods of science from other ways of knowing leaves science vulnerable to abuse and the public vulnerable to exploitation by all manner of fanatics and charlatans, ranging from creationists and scientologists to radical animal-rights activists and peddlers of new-age mysticism.

BSCS, in collaboration with the Social Sciences Education Consortium, in Boulder, recently completed a project supported by the National Science Foundation on teaching about the history and nature of science.[25] What follows is a sample of the suggested curricular themes from that report, with examples related to the translation of the molecular biology laboratory.

Principle 1. *The laboratory seldom provides definitive or final answers to scientific questions.* Although laboratory investigations in molecular biology may provide concrete data, it is a mistake to assume that those data always provide concrete and immutable answers to complex questions about life on earth. Uncertainty—resulting from indeterminacy[26] and "from the emergent properties of organisms at higher levels of biological organization"[27]—makes simple extrapolations from molecular data to complex characteristics difficult at best and dangerous at worst. In fact, much of the scientific objection to the HGP relates to the questionable ability of scientists to derive much helpful information about complex systems from an analysis of DNA sequences alone.[28] One hesitates to enter into a protracted debate about the relative merits of reductionism in the context of an educational program for high school students, but it seems only judicious that any translation of the molecular biology laboratory include a strong caveat about assuming that we understand life on earth simply because we have access to its constituent molecules. (This fallacious reasoning is behind some of the arguments to exclude dissection from the biology classroom.)[29]

Principle 2. *Scientific explanations are based on empirical observations or experiments.* Scientific inquiry assumes that the universe is explainable without appeals to supernatural phenomena. Observations are based on sensory experiences or extension of the senses through technology. Clearly, the molecular biology laboratory demonstrates this principle.

Principle 3. *Scientific explanations are tentative.* Explanations can and do change. There are no scientific truths in an absolute sense, and scientists often suspend final judgment on the answers to scientific questions. Molecular biology, in fact, has caused geneticists to revise their definition of a gene; the HGP may result in additional revisions.[30]

Principle 4. *Scientific explanations are probabilistic.* A statistical view of nature, not an absolute view, is fundamental to science. The probabilistic view of explanations is evident implicitly or explicitly when stating scientific predictions of phenomena or explaining the likelihood of events in actual situations. Geneticists long have faced the problem of conveying a statistical view of nature to the public. The identification of innumerable human genetic variations by molecular biologists will complicate the problem still further as we try to explain to the public the expression of those variations in different environments.

Principle 5. *Scientific explanations assume cause-effect relationships.* Cause-effect relationships are fundamental to making sense of phenomena, and much of science is directed toward determining causal relations and developing explanations for interactions and linkages between objects, organisms, and events. Distinctions among causality, correlation, coincidence, and contingency separate science from pseudo-science.

Principle 6. *Science cannot answer all questions.* Some questions simply are beyond the realm of science. Questions involving the meaning of life, ethics, and theology are examples of questions that science cannot answer.[31] Molecular biology, for example, might help us determine how *Homo sapiens* arose from its earliest mammalian ancestors, but it is powerless to determine why we are here, in the sense of final causes.

Principle 7. *Science is not authoritarian.* Garrett Hardin reminds us that "science is ineluctably married to doubt."[32] As *Science for All Americans* states, "No scientist, however famous or highly placed, is empowered to decide for other scientists what is true, for none are believed by other scientists to have special access to the truth."[33] Although the public may view disagreements among scientists as evidence that the science in question is somehow flawed (a favorite argument of creationists), disagreements and multiple competing hypotheses are essential to the health of the scientific enterprise.

That humans go into the laboratory in search of explanations for natural phenomena—and that we pursue knowledge even for its own sake—tells us something about what it is to be human. As far as we know, we are the only species that can look backward to its biological and intellectual history, and that can appropriate the future. Science, a uniquely human activity, illuminates both endeavors.

TEACHING AND LEARNING

Paul DeHart Hurd, of Stanford University, whom many regard as the most knowledgeable science educator in the world, noted in a 1991 speech that the most consistent attribute of the American educational system is amnesia. In successive educational reforms dating to the turn of the century, Hurd maintains, we have forgotten that given approaches did not work and therefore tried them again, to the detriment of students and teachers. In addition, we have forgotten, after brief experimentation, those strategies that do work, have failed to implement them broadly enough, or have neglected to nurture them in the ways that research shows are necessary if they are to endure. I suspect, therefore, that none of what I propose about teaching and learning science will sound new, because much of it has been proposed and tried before. Furthermore, most of these proposals have appeared in some three hundred reports published since the 1983 publication of *A Nation at Risk* and have been catalogued in numerous television and radio reports detailing the abysmal state of science education in this nation. Nonetheless, I shall reiterate three familiar injunctions about teaching and learning science in the hope that each of you will help them catch and stick somewhere.

Principle 1. *Science instruction should not subordinate a sound conceptual framework to a collection of disarticulated facts.* The translation of the molecular biology laboratory should occur in the context of instruction that emphasizes major themes in biology such as evolution, genetic continuity, and structure-function relationships. An empty taxonomy of terms in molecular biology without a central understanding of basic biological concepts is no more helpful than the empty phylogenetic taxonomy to which we subject students in the absence of a conceptual framework for evolution and ecology.

Principle 2. *Instruction in biology should expose students to the personal and social implications of progress in the biological sciences.* This symposium assumes that progress in molecular biology raises important issues for individuals, families, and society. Biology education should acquaint students with the scope of those issues and should help students develop the skills of rational inquiry necessary to assess the ethical and public-policy dimensions of the issues in a sound manner.[34] Students should understand that ethical analysis is a form of rational inquiry, as is science. Although there are few final, correct answers to such issues, students should be able to distinguish well-reasoned and badly reasoned arguments in support of competing positions.

Principle 3. *If we expect students to understand the nature and methods of science, they must spend much of their time doing science rather than reading about it.* Science education should reflect the essence of science, which is inquiry. That is best accomplished in the classroom when teachers eschew their traditional role of dispenser of information and instead guide students in the discovery and construction of their own knowledge.

Notwithstanding the pervasive philosophical agreement with the foregoing educational principles, their implementation has been less than widespread. Numerous barriers, ranging from institutional inertia and budgets to the influence of special-interest groups, preclude rapid and wholesale improvement in science education. Some of the barriers that impede the translation of molecular biology follow, along with brief descriptions of some practical solutions.

Barrier 1: An explosion of knowledge. The overwhelming amount of new information generated in molecular biology during the past two decades presents serious problems for those producing educational materials, as well as for teachers and students. In particular, we must beware the temptation to introduce new information into the curriculum simply because it *is* new. If we succumb, we might easily fail to integrate the new information into a cohesive picture of biology for our students.

The explosion of biological knowledge is not, of course, limited to genetics and molecular biology, so the precollege curriculum is confronted with the integration of vast amounts of new information from virtually all biological subdisciplines. The most difficult question for curriculum developers, therefore, is not "What do we put in?" but "What do we leave out?" Because there is no national consensus on the answer to this question, textbook publishers attempt to mention virtually every topic, lest their books be found wanting by state or local curriculum committees. The result is ever-larger textbooks that are long on facts and short on major concepts.[35]

Fortunately, two current efforts provide some hope of a resolution to the problem of the ever-expanding curriculum—and textbook. Project 2061, sponsored by the AAAS, is developing "benchmarks" for content at grades 2, 5, and 10. These benchmarks draw on the themes elaborated in *Science for All Americans* and provide concrete suggestions for the omission of much traditional content. Simultaneously, the National Research Council of the National Academy of Sciences and Engineering is developing national standards for content, teaching, and assessment in precollege science. These standards, which are scheduled to be in place by 1994, hold the promise of delimiting the content of the biology curriculum nationwide, a step that would help stem the unsatisfactory addition of new content for its own sake.

Barrier 2: Poor preparation of new teachers and failure to provide continuing education for experienced teachers. The undergraduate preparation of precollege biology teachers often leaves those nascent professionals with a disarticulated view of biology and with no practical exposure to research that conveys an understanding of the nature and methods of science. Furthermore, current biology teachers, many of whom are now in their forties or older, concluded their undergraduate preparation before the ascendance of molecular biology and have had limited opportunity to upgrade their skills and knowledge. During the last half-dozen years, the National Science Foundation (NSF) has supported a growing number of in-service education programs for biology teachers, and many of those programs focus on molecular biology. In addition, NSF funded in 1991 a study by the National Research Council to set standards for in-service programs in biology.

Several organizations and universities have developed materials and attendant in-service programs designed to improve the teaching of molecular biology. Among the most influential programs are those run by the North Carolina Biotechnology Center, Cold Spring Harbor Laboratory, San Francisco State University, the University of Kansas, and the National Association of Biology Teachers.

Barrier 3: A decline in the amount of time and money available for laboratory investigations. Science education should involve students in *doing* science by providing opportunities for authentic investigations in the laboratory and the field. Unfortunately, shrinking budgets for education continue to limit the dollars available for equipment and supplies, and administrative concerns about the crowded daily schedule often drive decisions about the curriculum, limiting the time available for science instruction in general and laboratory investigations in particular. Optimally, of course, the schedule should *reflect*, rather than *dominate*, decisions about the curriculum.

Other considerations impede a focus on laboratory investigations in the biology curriculum. For example, legal liability—in what may be the most litigious society on Earth—not only constrains trips to the field for observation and experimentation, but also limits the range of laboratory procedures available to precollege students. Increasingly vocal animal-rights activists—the most recent iteration of the antiscience movement—seek to abolish all uses of animals in American education, further limiting the options for genuine laboratory investigation.[36]

Numerous commercial and nonprofit groups have developed low-cost equipment and supplies to help introduce molecular biology into the precollege classroom. These materials include relatively inexpensive set-ups for restriction analysis, for transfection of plant cells by bacteria such as *Agrobacterium tumefaciens*, and for simple transformations of *E. coli*, using antibiotic-resistant plasmids. BSCS has developed a series of investigations that allow students to use colored paper clips to explore restrictive fragment length polymorphisms, to simulate the construction of recombinant plasmids, and to model DNA hybridization in the study of hominid phylogeny.

Barrier 4: A lingering perception that bioethics should not be part of the biology curriculum. A traditional view of science and science education holds that students should concentrate solely on the content and methods of science. More recent views emphasize the need for students to confront and analyze the personal and societal dimensions of science and technology.[37] This view reflects a strong consensus in the United States that precollege science instruction should prepare students to exercise their rights and duties as citizens in a democratic society where many of the public policy issues that confront the average citizen have a scientific or technological basis.

Fortunately, the rapid pace of progress in genetics and molecular biology, while confounding students, teachers, and curriculum developers with vast amounts of new information, also raises questions of great interest to young people, questions related to personal and community health, genetic screening, development of new pharmaceutical products and other biologicals, normality and abnormality, abortion, population control, genetic discrimination, substance abuse and addiction, mental health, and the history and future of life on the planet. There is much good curriculum that uses such issues as hooks to engage the students in biology and to emphasize the importance of understanding the relevant science before one can offer sound opinions on any of the issues at hand.

Barrier 5: A lack of training and experience among biology teachers in conducting effective analyses and discussions of bioethical dilemmas. Most high school biology teachers in the United States are trained in departments of biology in undergraduate colleges and universities, and they receive little preparation for teaching about bioethics and public policy. Once in the biology classroom, therefore, most teachers are uncomfortable conducting discussions and are unskilled in incorporating the components of effective ethical analysis into their instruction.

A number of programs have addressed these deficiencies during the last decade. For example, the Human Genetics and Bioethics Education Laboratory, at Ball State University, has trained hundreds of biology teachers to conduct ethical analysis in the classroom. In September 1992, BSCS, with support from the U.S. Department of Energy, distributed to all biology teachers in the United States a free module titled *Mapping and Sequencing the Human Genome: Science, Ethics, and Public Policy.* The hundred-page module includes twenty-five pages of background materials for the teacher about teaching ethics and public policy, and five class periods of instruction that involve students in ethical analysis and discussion of related public policy. The materials introduce students to a basic language of ethics and ask the students to consider the assumptions that define and describe molecular biology. Perhaps most important, the materials ask students to address assumptions about the extent to which DNA determines complex human traits and the extent to which the HGP can provide insights into the question, "What is it to be human?"

Barrier 6: Assessment of student learning that is not consistent with the objectives of conceptually based, inquiry-oriented teaching and learning. Multiple-choice tests that assess only the ability to recognize and recall trivial knowledge reinforce poor instruction and poor curriculum. If we wish students to demonstrate an understanding of major concepts and to apply those concepts to the solution of problems, the tests we use must reflect those objectives.

Barrier 7: Lack of involvement of the scientific community in the precollege science curriculum. American science education is in trouble partially because the scientific community has relinquished its responsibility for the integrity of the precollege science curriculum. Responsibility for the content and pedagogy of the biology curriculum, for example, now resides in the hands of a few publishers whose fact-laden but often inaccurate and conceptually empty books control 95 percent of the instructional time in the classroom.[38] Research scientists can contribute to the improvement of the curriculum in numerous ways:

Work with professional societies in science education and with the

National Academy of Sciences to help set new standards for science content and for teaching.

Work with publishers and groups such as BSCS on the design, writing, and review of new curricula.

Help to improve the undergraduate curriculum for prospective precollege science teachers, especially by ensuring that undergraduates have at least one rigorous exposure to research.

Work through their professional societies to establish summer programs that involve science teachers in research.

Volunteer to teach science in their local schools and get the necessary training from the teachers on how to teach effectively at the grade level in question.

Donate used equipment from their home institutions to local schools.

Be aware of the content of the local curriculum and offering help—not only criticism—when they find it wanting.

I am pleased to say that a number of the professional societies associated with molecular biology have been quite forthcoming in their support of precollege science. Among them are the American Society of Human Genetics, the American Society of Biochemistry and Molecular Biology, the National Society of Genetic Counselors, the American Society for Microbiology, the American Association of Immunologists, and the Society for Neuroscience. I am hopeful that still more professional societies will recognize the important role they can play and will become involved in the improvement of precollege biology.

THE FUTURE

Few American institutions are as resistant to change as the educational system, which resembles a complex ecosystem where the slightest perturbation in one component reverberates throughout. It is naive and inappropriate therefore to address the teaching of molecular biology in isolation, because any change in the content of the curriculum also affects budgets, books, tests, the education of teachers, and numerous other aspects of the system.

As knowledge of molecular biology expands, perhaps the greatest challenge to biology education will be the integration of that new knowledge into a cohesive picture of biology for the average student, much as the greatest challenge of the HGP will be the integration of the complete physical map of three billion bases into a cohesive understanding of the biology of *Homo sapiens*. The educational community cannot accomplish this integration of molecular biology into the curriculum by itself, nor should it. The scientific community must play a central role, working with educators to determine how best to make this difficult, sometimes frightening discipline understandable to the average person.

Because molecular biology will continue to raise difficult questions of ethics and public policy, scientists and educators alike must develop mechanisms for the rational consideration of those questions and must help develop in our students the skills and knowledge essential to informed, dispassionate analysis. Single-issue special-interest groups are likely to make such analysis increasingly difficult in the public school classroom, and scientists and educators must be willing to withstand the inevitable assaults from vocal zealots who oppose open discussion of views that conflict with their own.

The translation of science for the public never has been easy. We must overcome the inherent difficulty and foreignness of the information itself and, especially in biology, the anxiety that arises when new information forces us to confront and perhaps revise longstanding assumptions about what it is to be human.

As we set the criteria for success in programs that seek to enlighten the public about genes and human self-knowledge, we would do well to remember that 133 years after the publication of Darwin's *Origin of Species*, large numbers of individuals misinterpret and reject its assumptions and conclusions. Perhaps we have no right to expect that a mere dozen human generations are enough to permit wholesale acceptance of so revolutionary a framework as that proposed by Darwin. "The mind," it has been said, "likes a strange idea as little as the body likes a strange protein and resists it with similar energy."[39]

It is not clear that molecular biology will reveal insights as profound and potentially disturbing to the public as did Darwin's work. Neither is it clear, however, that we yet know all of the questions about molecular biology, let alone the answers. What is clear is the rapid pace of progress and the rate at which discoveries in basic science are translated into technologies that raise what once were intellectual abstractions to the level of difficult decisions for individuals, families, and society.

The stakes are quite high, and one hopes we will bring to the task of education the same level of talent, rigor, and resources that we devote to the science itself.

NOTES

1. Council of Chief State School Officers (CCSSO), *State Indicators of Science and Math 1990* (Washington, D.C.: CCSSO, 1990).
2. Biological Sciences Curriculum Study (BSCS), *Biology Teachers Handbook* (New York: John Wiley and Sons, 1978); Theodosius Dobzhansky, "Nothing in Biology Makes Sense except in Light of Evolution," *American Biology Teacher* 35 (1973): 125; F. James Rutherford and Andrew Ahlgren, *Science for All Americans* (New York: Oxford University Press, 1990); National Research Council (NRC), *Fulfilling the Promise: Biology Education in the Nation's Schools* (Washington, D.C.: National Academy Press, 1990).
3. Ernst Mayr, *The Growth of Biological Thought* (Cambridge, Mass.: Harvard/Belknap Press, 1982); and Mayr, *Toward a New Philosophy of Biology: Observations of an Evolutionist* (Cambridge, Mass.: Harvard University Press, 1988).
4. Barton Childs, "Science as a Way of Knowing: Human Genetics," in John A. Moore, ed., *Science as a Way of Knowing III—Genetics* (Baltimore: American Society of Zoologists, 1986), pp. 835–844.
5. Barton Childs et al., *Molecular Genetics in Medicine* (Baltimore: Johns Hopkins University Press, 1989).
6. Dorothy Nelkin and Laurence Tancredi, *Dangerous Diagnostics: The Social Power of Biological Information* (New York: Basic Books, 1989); Neil A. Holtzman, *Proceed with Caution: Predicting Genetic Risks in the Recombinant DNA Era* (Baltimore: Johns Hopkins University Press, 1989).
7. Charles R. Scriver and Barton Childs, eds., *Garrod's Inborn Factors in Disease* (New York: Oxford University Press, 1989).
8. Biological Sciences Curriculum Study (BSCS), *Curriculum Development for the Year 2000* (Colorado Springs: BSCS, 1989); Joseph D. Novak and D. Bob Gowin, *Learning How to Learn* (Cambridge: Cambridge University Press, 1984).
9. John A. Moore, "Science as a Way of Knowing: Evolutionary Biology," in John A. Moore, ed., *Science as a Way of Knowing I—Evolutionary Biology* (Baltimore: American Society of Zoologists, 1984), pp. 467–534.
10. Holtzman, *Proceed with Caution*; Neil A. Holtzman and Mark A. Rothstein, "Invited Editorial: Eugenics and Genetic Discrimination," *American Journal of Human Genetics* 50 (1992): 457–459; Marvin R. Natowicz, Jane K. Alper, and Joseph S. Alper, "Genetic Discrimination and the Law," *American Journal of Human Genetics* 50 (1992): 465–475.

11. Joseph D. McInerney, "Biotechnology in Perspective: Public Education," *Biotechnology Education* 2 (1991): 98.

12. Niles Eldredge, *Time Frames: The Evolution of Punctuated Equilibria* (Princeton, N.J.: Princeton University Press, 1985).

13. Robert Boyd and Peter Richerson, *Culture and the Evolutionary Process* (Chicago: University of Chicago Press, 1985); Luigi L. Cavalli-Sforza, "Cultural Transmission and Nutrition," in Artemis P. Simopoulos and Barton Childs, eds., *Genetic Variation and Nutrition* (Basel: Karger Press, 1990).

14. Childs, "Science as a Way of Knowing"; Biological Sciences Curriculum Study (BSCS), *Evolution: Inquiries into Biology and Earth Science* (Seattle: Videodiscovery, Inc., 1992).

15. BSCS, *Evolution*.

16. Denis P. Burkitt and S. Boyd Eaton, "Putting the Wrong Fuel in the Tank," *Nutrition* 5 (1989): 189–191; S. Boyd Eaton and Dorothy A. Nelson, "Calcium in Evolutionary Perspective," *American Journal of Clinical Nutrition* 54, no. 1 (1991): 281S–287S.

17. Burkitt and Eaton, "Putting the Wrong Fuel in the Tank."

18. Biological Sciences Curriculum Study (BSCS) and the American Medical Association, *Mapping and Sequencing the Human Genome: Science, Ethics, and Public Policy* (Colorado Springs: BSCS, 1992), p. 10.

19. James Scott, "Molecular Genetics of Common Diseases," *British Medical Journal* 295 (1987): 769; Thorkild I. A. Sorensen et al., "Genetic and Environmental Influences on Premature Death in Adoptees," *New England Journal of Medicine* 318 (1988): 727–732; Robert Plomin, "The Role of Inheritance in Behavior," *Science* 248 (1990): 183.

20. Jean W. MacCluer and Candace M. Kammerer, "Invited Editorial: Dissecting the Genetic Contribution to Coronary Heart Disease," *American Journal of Human Genetics* 49 (1991): 1139; Theodore Friedmann, "The Human Genome: Some Implications of 'Reverse Genetic' Medicine," *American Journal of Human Genetics* 46 (1990): 407; Holtzman, *Proceed with Caution*.

21. Patricia A. Baird, "Genetics and Health Care: A Paradigm Shift," *Perspectives in Biology and Medicine* 33 (Winter 1990): 203–213; Childs, "Science as a Way of Knowing"; Holtzman, *Proceed with Caution*; George C. Williams and Randolph M. Nesse, "The Dawn of Darwinian Medicine," *Quarterly Review of Biology* 66 (1991): 1.

22. Rutherford and Ahlgren, *Science for All Americans*.

23. National Center for Improving Science Education (NCISE), *Science and Technology for the Elementary Years: A Curriculum Framework* (Washington, D.C.: NCISE, 1989).

24. Moore, "Science as a Way of Knowing: Evolutionary Biology."

25. Biological Sciences Curriculum Study (BSCS) and Social Science Education Consortium (SSEC), *Teaching about the History and Nature of Science and Technology: A Curriculum Framework* (Colorado Springs: BSCS, 1992).

26. Mayr, *Toward a New Philosophy of Biology*.

27. Marsden S. Blois, "Medicine and the Nature of Vertical Reasoning," *New England Journal of Medicine* 318 (1988): 847.

28. Bernard D. Davis, ed., *The Genetic Revolution: Scientific Prospects and Public Perceptions* (Baltimore: Johns Hopkins University Press, 1991); Richard C. Lewontin, "Book Reviews: The Dream of the Human Genome," *New York Review of Books* (May 28, 1992): 31–40; Alfred I. Tauber and Sahotra Sarkar, "The Human Genome Project: Has Blind Reductionism Gone too Far?" *Perspectives in Biology and Medicine* 35 (1992): 220.

29. Joseph D. McInerney, "Oppose PETA School Campaign," letter to the editor, *American Biology Teacher* 54 (1992): 70; Adrian Morrison, "Speciesism: A Perversion of Biology, Not a Principle," letter to the editor, *American Biology Teacher* 54 (1992): 134–136.

30. Elof A. Carlson, "Defining the Gene: An Evolving Concept," *American Journal of Human Genetics* 49 (1991): 475.

31. Stephen J. Gould, "Impeaching a Self-Appointed Judge," book review, *Scientific American* 267 (July 1992): 118–121.

32. Garrett Hardin, "Human Ecology: The Subversive, Conservative Science," *American Zoologist* 25 (1985): 469.

33. Rutherford and Ahlgren, *Science for All Americans*.

34. NRC, *Fulfilling the Promise*; Rutherford and Ahlgren, *Science for All Americans*.

35. Harriet Tyson-Bernstein, *A Conspiracy of Good Intentions: America's Textbook Fiasco* (Washington, D.C.: Council on Basic Education, 1988).

36. McInerney, "Oppose PETA School Campaign."

37. NRC, *Fulfilling the Promise*; Rutherford and Ahlgren, *Science for All Americans*.

38. Connie Muther, "What Every Textbook Evaluator Should Know," *Educational Leadership* 42 (1985): 4–8.

39. Wilfred Trotter, "Quotation in the Conclusion," from *Perspectives in Biology and Medicine* 34 (1991): 549.

Communicating Genetics: Journalists' Role in Helping the Public Understand Genetics

Larry Thompson

A substantial majority of Americans do not have a sufficient vocabulary or comprehension of concepts to utilize a wide array of scientific communication.—Jon D. Miller, *Washington Post*, June 2, 1986

The public . . . can assimilate an astonishing amount of technical information if they feel that it is necessary to protect themselves in a dispute.—Robert C. Forney, *Christian Science Monitor*, September 25, 1986

Public opinion in this country is everything.—Abraham Lincoln, September 16, 1859

All three statements, taken from a report by the Office of Technology Assessment, a congressional agency that studies the impact of science and medicine on the lives of Americans, are true. Overall, Americans know woefully little about science, despite its importance in their daily lives. High school biology and chemistry seem to be for eggheads only. Computer classes have yet to become universal, and science requirements in college remain minimal. Worse, Americans do not seem to care much about why they do not understand. That leaves the general media—newspapers, magazines, and television—as the principal sources for the public's understanding about one of the major intellectual forces that continuously reshapes society.

Are the media doing a good job? Can they do the job? While recognizing their power, the media collectively resist taking on the role of teacher. Consequently, there are no easy answers. There are no good surveys to measure the media's effectiveness. Yet, from a smattering of sources, a somewhat reliable picture can be constructed. Just as there is no one "science," and there is no one "medium," there is no one "public." The public is a widely diverse collection of many different groups—racial, religious, ethnic, educational—that includes the less-diverse communities of scientists and the media.

Assessing public perceptions through random population surveys is like painting by numbers. What you get is a picture that depicts a duck, but it is produced with block colors that depend on strict lines between them and do not reflect the shadings and blendings of colors that make up the real world. Yet, you can still tell it's a duck, so painting popular attitudes by statistical assessments can be a useful exercise, even if it is imprecise.

HEALTH CONCERNS AND ATTITUDES

My personal belief is that everyone has some interest in medicine. Everyone has a body. Most people would like to understand at least a little about how it works. And they become very attentive—in the jargon of social scientists, they become "information seekers"—when something goes wrong, when they or family members get sick.

The American Cancer Society conducted a random survey of 1,045 people in August 1991 and found that "a majority of the respondents reported awareness of recent information about each of the health care issues surveyed, particularly AIDS, cancer and drug addiction. Awareness . . . tended to be related to the level of education, with more educated individuals more often expressing awareness of these issues."[1]

But when asked which were the worst health problems facing the nation, the public got it wrong:

31 percent said AIDS was the worst health problem (far greater than the percentage of deaths it actually causes).

28 percent said cancer (about proportionate to the number of deaths actually caused by this disease).

15 percent said heart disease (much too low when compared to the number of deaths).

6 percent said drug addiction.

2 percent said obesity.[2]

The Public Health Service's national health statistics show that heart disease, by far, is the worst health problem. It kills more than a third of everyone who dies—both men and women, and not just men, as is commonly assumed—in America. That is more than half a million people per year. By contrast, from the time the federal Centers for Disease Control in Atlanta, Georgia, began keeping AIDS data in 1981 until February 1992, the total number of cases had reached 213,641, of which 210,043 were adults. Of all persons diagnosed with AIDS, 138,195 people had died. That is a drop in the bucket compared to heart disease and cancer, but because AIDS has received so much press attention during the last decade, it weighs more heavily on the public mind. The same can be said for the dangers of drug addiction, which does not kill as many people as cancer or heart disease, but clearly gets much more media attention.

When asked in a *Washington Post* national survey whether the government should increase funding for AIDS research, 41 percent of those questioned said they wanted it increased "a great deal" and 25 percent wanted it increased "somewhat." For cancer, 44 percent wanted research funding to increase a great deal and 25 percent wanted it to increase somewhat. For many other programs, from arts and music (7 percent increase a great deal; 11 percent increase somewhat) to public TV and radio (6 and 13 percent), to the space program (12 and 16 percent), to food stamps (9 and 12 percent), to the military (7 and 11 percent), support for spending more money was lukewarm.[3] The statistics indicate individuals' concerns about their own health and their desire for someone else to spend the money necessary to protect it.

Yet when the *Los Angeles Times* conducted a survey of nearly two thousand American adults on public attitudes about personal health in the summer of 1989, it found some surprising information as to what motivates Americans about healthy behavior and seeking health information. It fits my "everyone has a body" theory. When asked what they wanted out of life, 43 percent of the men and 55 percent of the women said they wanted to be healthy. The next biggest response category was to help others: 16 percent of the men and 17 percent of the women. Marriage was lower on the wish list: 13 percent of the men wanted to be married; only 8 percent of the women did. Only 9 percent said they wanted to be rich; only 2 percent wanted to be powerful.[4]

A majority of those surveyed turned their desire to be healthy into action: 60 percent of the men and 53 percent of the women said they exercised regularly. Of those, 67 percent of the men and 71 percent of the women said they did it to stay healthy. Other healthy behaviors in the previous year included:

Started dieting, 28 percent.

Launched an exercise program, 21 percent.

Stopped smoking, 11 percent.

Stopped drinking, 8 percent.[5]

Of course, these numbers are not great, but they show a regular level of healthy actions in at least a quarter of the population. They are motivated to "do something," to change their behavior to improve their personal health.

Still, health messages are not getting through to the hard-core, sedentary junk-food eaters who still smoke. One-third of the population continues to smoke cigarettes despite the clear risk of cancer and heart disease. A recent National Cancer Institute report concluded that Americans are not eating the recommended portions of vegetables every day. More than 40 percent of the population does not exercise at all. And the President's Council on Fitness and Sports concluded a few years ago that our kids are out of shape. We simply fail to teach them to be healthy.

Yet it is clear that the media can have an important impact on health behavior, especially when the media make a fuss about something. This could be clearly seen in 1990 when the National Cancer Institute (NCI) and the Jacobs Institute of Women's Health published their joint Mammography Attitudes and Usage Study. In 1989, NCI had been making a push to get women over forty to have a mammogram on a regular basis. The media—print and broadcast—picked up on the issue and produced a number of stories throughout that year.

The NCI and the Jacobs Institute then randomly surveyed 980 women, forty or older, across the country and found that two-thirds of the women in the at-risk groups had had a mammogram, a substantial increase over the number of women (only about one-third) found to have had one by the National Health Interview Survey in 1987. Still, only one-third were getting follow-up mammograms on the schedule recommended by the NCI. So while a message was getting out, the response was mixed. And what were the greatest predictors of who would get a mammogram? This is not a surprise. The most likely to get a mammogram were white, had a household income at or above fifty thousand dollars, had a college degree or higher education, and were married.[6]

GENERAL SCIENCE CONCERNS AND ATTITUDES

Each year the National Science Foundation (NSF), a branch of the federal government similar to the National Institutes of Health, hires the Public Opinion Laboratory of Northern Illinois University to conduct an ongoing series of studies on public attitudes about science. The survey of 1990, the latest year for which results are available, questioned 2,033 adults nationwide by phone.

The NSF survey reached a simple though distressing conclusion: "Most Americans are not familiar with fundamental scientific theories and do not understand basic characteristics of simple technologies."[7] Yet, the survey found that even though knowledge is low, interest is high. But again, it is a mixed picture.

Only 8 percent of the population, some fourteen million Americans, are "attentive" to new scientific discoveries. "Attentive" refers to those who actively and regularly seek out information about science, medicine, and technology. The level of interest varies by topic, from 6 percent attentive to new advances in space exploration to 20 percent interested in environmental pollution.

In addition to those who are attentive and actively seek out information, a larger portion of the population says it is interested in new scientific information. The 1990 NSF survey found that 39 percent of the adult population described itself as very interested in new scientific discoveries. About 40 percent were interested in new inventions and technologies. Nearly 70 percent are very interested in new medical discoveries. And about 65 percent are very interested in environmental pollution.

Interest, however, does not necessarily translate into knowledge. More than one-third felt they were poorly informed about science or medicine. Most judged their knowledge on issues of science and medicine to be only moderate.

As part of the Public Opinion Laboratory survey, thirteen questions were formulated on and around various scientific statements, such as "The center of the Earth is very hot"; "Lasers work by focusing sound waves"; and "Which travels faster: light or sound?" The results were distressing: Only 13 percent of the population could answer at least half the questions correctly; only 1 percent got them all right. Seventy percent did not know that antibiotics do not kill viruses.

The implication of the survey is that a substantial portion of the population does not care about science. However, as Carol Rogers points out, while people may not have much general scientific information, they become incredibly knowledgeable when they have a good reason to do so: when they get cancer, or their child becomes sick, or a nuclear power plant or garbage incinerator is about to be opened nearby.[8] Personal motivation or personal risk makes people become science-information seekers. Frequently, that information-seeking behavior is driven by the media, and the news stories are driven by novelty and controversy. The more controversial a scientific issue becomes, the more information the media put out and the more the public pays attention and learns about it.

OFFICE OF TECHNOLOGY ASSESSMENT GENETICS SURVEY

In an effort to understand what the public knows and feels about genetics and biotechnology, the Office of Technology Assessment (OTA) hired the polling firm of Louis Harris and Associates to conduct a representative survey of national attitudes. From October 30 to November 17, 1986, Louis Harris pollsters telephoned 1,273 Americans about their attitudes.[9] While this survey was conducted several years ago, I believe that it still accurately reflects public perceptions because—despite the advent of human gene therapy and the discovery of several genes that cause serious illnesses—there has been no cataclysmic event to draw unusual attention to the field of genetics. (A cataclysm might include the death of a gene therapy patient, or the production of a cancer by gene therapy, or some other highly visible and newsworthy event, perhaps related to the Human Genome Project.) I also don't believe the NIH's move to patent a large number of genes (now more than three thousand) has generated enough public debate to alter the survey's findings.

The OTA survey had some expected results and some surprises. The survey asked people to rate their own basic understanding of science and technology: 16 percent said very good; 54 percent said adequate; and 28 percent indicated that their scientific understanding was poor. The OTA claims this level of basic understanding is going down; the NSF says it has been stable over the last decade.

The OTA survey found "interest" in science and technology at only 23 percent, substantially smaller than the 41 percent the NSF found in 1985. The OTA concluded that only 20 percent of the population was "scientifically attentive." The basic level of scientific attentiveness was less than indicated in the NSF-funded surveys. Only one-third said they were "very concerned" about government policy concerning science.[10]

But when these groups are combined, nearly half (47 percent) of the adult population described themselves as "very interested, very concerned or very knowledgeable about science and technology." These are the persons that Jon Miller, the director of the NSF survey, calls "scientifically attentive" or are, as the OTA relabeled them, "scientifically observant."

Attitudes and Concerns about Genetics:
The OTA Report

Even if the public does not really know much about what is going on in science, 80 percent of the public expects benefits from science and technology in the next twenty years. However, 71 percent believe that science will present some risks to themselves and their families. Although only 19 percent say they have heard about any potential dangers from genetically engineered products, 52 percent believe that genetically engineered products are at least somewhat likely to represent a serious danger to people or to the environment.

Other beliefs about genetics include the following: 61 percent believe that genetic engineering will produce antibiotic-resistant disease; 57 percent believe it will produce birth defects; 56 percent think it will create herbicide-resistant weeds; and 52 percent fear genetics will endanger the food supply. Even with those concerns, 62 percent of the public believe the benefits of genetic engineering outweigh the risks. A hard-core minority, 28 percent, think the risks are too great.

The public's ability to sort out risk numbers and make a rational choice is highly questionable, since the results from the survey are somewhat contradictory. Over half of the persons surveyed (55 percent) said they would use a genetically produced organism that significantly increases farm production even if the risk of losing some local species were 1 in 1,000, a high risk of environmental impact. But the public is unwilling to accept unknown risks: 55 percent would not approve use of a gene product with "unknown but very remote" risks compared to one with a known risk of 1 in 1,000.[11] When the public is uncertain of the dangers, it tends to reject the technology, or the choice.

Attitudes about Human Gene Therapy: The OTA Report

Manipulating genes in humans has the potential to be one of the most sensitive and emotionally loaded issues of all technology. But public views, according to Louis Harris, are in conflict. While 52 percent believe it is not morally wrong to change the genetic makeup of human cells, 42 percent say it is morally wrong. Those are the general views when the question is asked in the abstract. But when the questioning gets specific and related to individuals, especially children, the public view changes:

84 percent of the public would strongly approve or somewhat approve using gene therapy to "stop children from inheriting a usually fatal genetic disease." Such action would be considered germ-line therapy, in which the sperm and egg cells would be genetically altered. This type of manipulation is so controversial among scientists that it is not now under consideration.

83 percent of the public would approve using gene therapy to "cure a usually fatal genetic disease." This would include the first effort at gene therapy treatment, provided to two children with severe combined immune deficiency who have been treated at the National Institutes of Health since September 1990.

77 percent would approve using gene therapy to "stop children from inheriting a nonfatal birth defect."

77 percent would use it to "reduce the risk of developing a fatal disease later in life," such as Huntington disease or even heart disease or cancer.

A substantial portion of the public would accept the use of genetic manipulation to carry out treatments that scientists themselves—for the most part—do not now consider acceptable. For example, 44 percent of the public would allow the use of genetic engineering to improve the intelligence a child inherits, and 44 percent would use it to improve inherited physical characteristics.[12]

The public attitude about germ-line therapy was a surprise. The OTA survey shows that the public is not as concerned about the difference between germ-line and somatic-cell gene therapy as is the scientific community. Somatic-cell therapy is any treatment where genes are inserted into the body of a single patient, and it affects only that patient. Ethicists have concluded that somatic-cell gene therapy is no different from drug treatment or surgery. Germ-line therapy, however, permanently alters the sperm or eggs of the treated individual, so that the changes are inherited by all future generations.

The OTA asked the following questions: "Suppose someone had a genetic defect that would cause a usually fatal disease in them, and it would likely be inherited by their children. Should doctors be allowed to correct the gene in the adult patient, correct only the gene that would carry the disease to future generations, both, or neither?" The results:

62 percent of the public said both—it should be genetically corrected in the adult patient and in such a way that it corrects the defect for future generations.

8 percent said it should be done in the patient only.

14 percent said in the offspring only.

11 percent said neither.[13]

Then the OTA survey became personal and asked whether people would be willing to undergo genetic treatment themselves. Of those questioned, 78 percent said they would undergo human gene therapy to correct a genetic proclivity to a serious or fatal disease. A few more, 86 percent, said they would do it for their kids (suggesting that at least 8 percent would not do it for themselves, but would do it for their children).[14]

WHAT SHAPES THE PUBLIC'S PERCEPTIONS ABOUT GENETICS?

There is no question that the media can influence what the public thinks about science. This becomes especially true when a scientific issue—such as human gene therapy—becomes controversial. Public attitudes about science and genetics, however, are shaped by many other, often conflicting, forces, including the following:

Level of education. Rae Goodell at the Massachusetts Institute of Technology found that taking at least one science course in college was a good predictor of future scientific interest.

Religious socialization and values. Years of indoctrination often shape a person's moral views sufficiently that they do not change with scientific information. The OTA survey found that those most likely to oppose genetic intervention, even to save the life of a child, tended to say they were very religious. Moreover, a quarter of the surveyed population thought it was morally wrong to create hybrids of plants and animals through genetic manipulation, either by genetic engineering or by the classic biological techniques of cross-fertilizing plants or crossbreeding animals. This group was distinguished by having both a lower level of education and a greater sense of religiousness.

The National Science Foundation survey found that 41 percent of those surveyed got the question on human evolution wrong not because they did not know the answer, but because they disagreed with it. Instead, they held to the viewpoint that God created humans, not that humans evolved from apelike ancestors. In the NSF survey, 14 percent said they did not know whether theories of human evolution are true or false. Altogether, 55 percent of the persons surveyed either rejected outright or were uncertain about the theory of evolution. A similar conflict showed up when they were asked whether the universe was created in a huge explosion: 33 percent said no, and 35 percent said they did not know.[15] This theory of cosmological origins contradicts literal interpretations of the biblical story of creation.

Family attitudes. Obviously, if you grow up in a family that is interested in science, you will tend to be more interested in science than if you grow up in a family that is indifferent to or uninterested in science.

The power of stories: books, TV, and movies. Ever since Shelley's monster graced the pages of her book about Frankenstein, mad scientists have been a common theme in science fiction. The portrayal of scientists as dangerous people who do dangerous things has had a profound impact on the public perception of research. That view reached its pinnacle during the 1950s, when crazy scientists, often using radiation, were portrayed as threatening the world with giant mutated insects.

Studies in the 1970s and 1980s at the Annenberg School of Communication at the University of Pennsylvania showed a significant change. The Penn study found that the public attitude had changed, mostly because of TV shows. Scientists were not seen as crazy and out of control, but rather were seen as "mean and scary." They did things for greedy reasons, for power, or just because they could do them. In the process, they created problems for humankind. That TV image carried through in the news pages where readers could see lots of evidence for scientific threats: Love Canal, the ozone hole, the green-house effect, nuclear power, and so on. The Penn study correlated heavy TV watching with a more negative view of science.

The news media. The media include any broadcast medium or pub-lication, both local and national newspapers, trade journals, maga-zines, the Big Three network news shows, and CNN. It is essential to realize that "the media" do not form a monolithic group. In a modern society swamped with information, individuals tend to specialize in what they are interested in—science, economics, foreign policy, inte-rior decorating. People get information from many different sources, and use no one source consistently. Television, not surprisingly, gets the most attention, according to the NSF survey:

According to the poll, 75 percent of the public watch TV regularly; 22 percent occasionally; and 4 percent not at all.

Only 57 percent read newspapers every day; 24 percent a few times a week; 8 percent once a week; and 10 percent less than once a week. Daily newspaper readership has declined from 69 percent of the popu-lation in 1973 to 53 percent of the population in 1990.

In addition, 23 percent read the news magazines regularly; 13 per-cent occasionally; and 64 percent never.[16]

The OTA study found similar patterns:

25 percent of the public report reading books or magazines on sci-ence and technology daily; 6 percent weekly; and 19 percent monthly.

36 percent say they read science sections in newspapers.[17]

WHAT MAKES NEWS?

If the media are the principal source of news about science and technology, then what is news? Who decides, using what criteria?

First I need to point out that newsroom leadership views science very much as does the general public. A small percentage of reporters and editors understand science, a smaller number, probably, than in the general population, because studies that survey the general popu-lation also include scientists themselves. Newsrooms, by their very nature, tend to exclude scientists. The consequence of that is not trivial. Science is seen as an important part of our world and it is judged to be worth covering, just as are politics, economics, movies, art, and music. But science is not as well understood in the newsroom as is politics or music. Science is not visceral. It is intimidating to these otherwise bright individuals. Consequently, science tends to be minimized, either relegated to stories in science sections, or limited to "breakthrough" stories (a term never used by true science writers, since almost nothing is a breakthrough), or covered as stories about the politics or economics of science. In a very thoughtful three-year study of science writers, published in *Health Forum,* Jay Winsten at the Harvard School of Public Health called science reporters the backbenchers of big-time journalism.[18] They are seen as masters of arcane subjects that only occasionally merit page-one attention or prime-time television coverage.

Political Controversies

This is the stuff that fuels news stories and captures public attention the most. And this is what the media do best. We love controversy. It puts edge in the story, it portrays the battle between good and evil. Just consider the recent debates about the appropriateness and safety of breast implants. There were several hundred stories, in all media outlets. There was the initial failure of governmental action, then the aggressive Food and Drug Administration action. There were anguished women with implants, and those women who wanted implants. There was a company that appeared to have failed to carry out its responsibilities. There were deadlines and decisions. The deadlines were reached, the decisions made, and now the story is out of the news, at least for the moment.

In genetics, the most important public debates occurred in the 1970s when scientists, the Congress, the public, and the media wrestled with the questions raised by recombinant DNA and the techniques used to splice bits of genetic material into novel combinations. Scientists raised questions about the dangers of creating super-strains of dangerous infectious organisms. Their concerns caught the media's and the public's attention. It took more than five years of clashes to conclude that the risks were not as high as first thought. Eventually, the concerns died down and so did media coverage.

MIT's Rae Goodell analyzed the recombinant debates, how they played out in the media, and how they affected the public. The whole debate was generated by the scientific community itself. Its leaders brought it to the attention of some of the media, yet it leaned on them—including the *New York Times*—to hold off on stories for some period of time. This kind of pressure still goes on. Everyone knows the names of the little girls who underwent the first human gene therapy, but no one, including the *New York Times* and the *Washington Post*, has published them.

The organizers of the Asilomar conference, at which the recombinant-DNA concerns were first publicly discussed, initially planned to ban journalists. Stuart Auerbach of the *Washington Post* threatened to file a Freedom of Information Act request since the conference was funded by NIH, a federal agency. The scientific leaders relented and invited fifteen journalists, but the ground rules said no stories could be written until the conference was over. As a result, the fifteen journalists reflected only the view of the scientists at the meeting. No scientists not invited to the conference were called for opinions. The Asilomar view, a consensus, was presented as the story. The stories had more depth than usual, but they lacked a broader look at all the related issues.

As the debates went on, the Congress was roused, and leading scientists tried to quell the increasing attention, first by disparaging scientists with contrary views and then by refusing to talk to reporters who did not always agree with the Asilomar scientists. These strategies did limit the stories to some degree. Eventually, the recombinant DNA story, like all stories, wound down, and reporters lost interest.

Behavior by Science Writers

Science writers, like any specialty writers in any newsroom of any kind of publication or broadcast medium, tend to have certain patterns of behavior that influence their reports. Most of these behaviors are not intended to bias their conclusions, but I think an argument can be made that they do. For example, science writers tend to accept the values of the establishment scientific institutions such as the National Academy of Science and the National Institutes of Health. As a result, their stories tend to reflect the norms and the interests of the scientific community. Similar relationships can be seen between business writers and large companies, sports writers and the teams they cover, and even between adversarial political writers and the politicians they cover. This common pattern occurs, in part, because reporters need to rely on ongoing relationships with their sources, whether in the scientific community or in the White House.

Some critics of journalism complain that science journalists rely too much on establishment scientific sources. The critics say that science reporters are lazy, that their stories rely only on the weekly publication of a few leading scientific journals, and that they do not produce enough novel stories. That might be true. Science reporters do tend to rely on scientific journals, meetings, press conferences, and organized publications produced by the scientific community. This very process, however, determines who is scientifically legitimate—and that influences who gets covered and what news gets out.[19]

How a Newsroom Works

The leaders in the newsroom have about as much scientific interest and literacy as the rest of the general population—sometimes, it seems, even less. To measure the general interest of newsroom leaders, look at what is on the front page. Politics. Politics. Politics. Economics. Politics. Race relations. Politics. Crime. Occasionally science, medicine, art, or dance. In my experience, general-publication editors are not that interested in science. They are dazzled by the grandeur of landing on the moon, or the horror of the shuttle blowing up. One study found that 90 percent of the population saw the tape of the Challenger exploding within twenty-four hours of the accident. But the attraction is the terrible beauty of a catastrophe, a train wreck from which you cannot take your eyes. The attraction and interest does not have to do with the creativeness of science, or the quest for understanding, or even a cure for a disease. Telling that kind of story is much more difficult.

A newsroom is characterized by competition. It is a place filled with bright, ambitious people. They all want to have the best and sexiest stories to dazzle their editor. There is competition for the front page. It's the only place to be for a journalist. The best journalists have an instinct for what makes a page-one story. That page can only hold so many stories, usually four to eight. The result is that a science writer has to compete with the reporters in the capital, the Congress, the financial centers, sports—every department in the place.

The study by Harvard's Jay Winsten found that this competition within the newsroom tended to distort science stories. Reporters try to make their story as strong and as significant as they can, so they can get on page one. Sometimes they go too far. I have read hundreds of stories about scientific advances that would take researchers one step closer to a cure for cancer, AIDS, fill-in-the-blank diseases. Every one of those characterizations was probably true. Yet neither cancer nor AIDS has been cured. Were the stories hyped? Probably.

Yet to some degree this hyperbole is understandable. Scientific concepts are hard and attention spans are brief. To get a reader—an editor—to pay attention to a story, science writers often have to water down the technical details, the science, and focus on the study's importance—a cure for cancer. One long-time science writer, Victor Cohn of the *Washington Post*, says there are only two kinds of medicine (or science) stories: New Hope or No Hope. Cohn jokes about it, but it's true. Editors really care only if a story says that there is new hope of curing some miserable disease because of the latest advance, or, often in a feature about a single individual with whom the reader can identify, that there is no hope.

In defense of science writers, it must be said that they have one of the hardest jobs in the newsroom. They deal typically with complex, arcane topics of immense technical difficulty. They must first find an interesting story that meets the basic criteria of what qualifies as news: it's new, has significance, and affects readers' lives. Then they must translate the science into a language that a general audience with grade-school comprehension can understand. Metaphors and analogies abound. Sometimes they become misleading.

Obstacles to Accurate Reporting and Writing

In addition to the more intellectual struggle over science stories in the newsroom, all reporters and writers face a constellation of concrete problems that affect every kind of writer, from those covering science to those reporting on the Congress. They include the following:

Lack of time to talk to enough people to really understand what is going on. Reporters are often like waiters. They skip from story to story as quickly as a waiter moves from table to table delivering orders. The result is that the writer is never an expert on the topic. That makes evaluating stories difficult. The reporter is forced to rely on the experts.

Lack of space to explain adequately the issues involved in the depth needed. Almost all stories are better when they are short. That accommodates the commercial-length attention span of most readers and viewers, but sometimes a story is so complex that it takes a longer time to tell it. Getting that time—or space—is often difficult.

The chilling effect of the so-called Ingelfinger rule and the *New England Journal of Medicine*. Several years ago an editor of the *NEJM* decided that the journal would not publish articles that had already been covered or mentioned in the press. As a result, many scientists refuse to describe their work until it is published in a peer-reviewed journal. They fear that if their results are first described in a general-circulation publication, the *NEJM* or *Nature* will refuse to publish it. Thus a handful of elite journals virtually controls the flow of scientific information.

The battle for influence. Everyone is trying to influence journalists. Public-relations firms flood journalists with information, trying to get them to do stories on their clients. Some scientists present themselves for credit for a discovery, or attempt to control the spin of a story. They can do spin control in many ways: by launching a story favorable to themselves, by providing contrasting information, or by refusing to talk about what is going on. The newsroom is in the eye of the storm. It is the vortex, the funnel, through which so much information, so many ideas seeking expression, flow. Reporters spend a great deal of their time just reading the ton of mail they get each day. They also waste a great deal of time chasing down leads from wires or tips or phone calls that never become stories.

Reporters and editors burn out. All stories lose their interest and appeal eventually. It has happened to AIDS, the War on Cancer, and even the MIAs of Vietnam. Some reporters are always willing to write about each of these issues, but the early torrent of stories that poured out of computers around the world on these topics eventually and inevitably slowed.

What does all this mean? The surveys clearly show that there is a reasonably high level of reader interest in science, especially in issues related to personal health and fitness. But the government studies by the NSF and the OTA show that the level of scientific knowledge among the general population is pretty low, given the importance of technology to American society. Moreover, both surveys show that about half the population is simply uninterested in science and technology. That makes it less likely that the general population will participate in thoughtful debates about the application of new technologies unless, for example, they perceive that a threatening plant is about to be built next to their house.

The failure of the general population to perceive any problems with the application of germ-line gene treatments—a topic so touchy that scientists are reluctant to discuss it, let alone do it—reinforces the idea that most Americans do not understand genetics. In conversations about science stories, I often find myself having to start at the beginning by describing what a gene actually is and does.

The general-circulation media are not going to solve this problem for society. Rather, the media have the same scientific malaise as the rest of the society. There are plenty of problems in trying to get science stories into newspapers, and hit-or-miss journalistic coverage is not going to redress the basic failure of the education system to make science compelling enough to keep the public attentive.

Improved public education may be the only way to improve scientific literacy and attentiveness to genetic issues. Studies clearly show that the higher the education level, the higher the individual's interest in science and medicine. (Perhaps most important is the taking of a college-level science course.) Given the pluralistic values of American culture and the expectation that every individual will pursue his or her own interests, I doubt that there will be any dramatic shifts anytime soon in the level of scientific attentiveness, whether the media write more science stories or not.

NOTES

1. American Cancer Society Poll by the Gallup Organization, Inc., August 1991, about public's attitudes about medicine and cancer research.

2. Ibid.

3. *Washington Post*/ABC News Poll, 1989, including questions about government funding for medical research.

4. *Los Angeles Times* Poll, no. 191, for the six days ending July 13, 1989, including questions on attitudes about personal health.

5. Ibid.

6. "Use of Mammography—United States, 1990," *Morbidity and Mortality Weekly Review* (*MMWR*) 39 (1990): 621–630.

7. National Science Foundation, *Science and Engineering Indicators*, 10th ed. (Washington, D.C.: NSF, 1991). Includes Public Opinion Laboratory of Northern Illinois University survey on public understanding of science.

8. Interview, Carol Rogers, former director of communications, American Association for the Advancement of Science, and science historian, University of Maryland, College Park, April 1, 1992.

9. Office of Technology Assessment, *New Developments in Biotechnology. Two: Background Paper, Perceptions of Biotechnology* (Washington, D.C.: OTA, 1987). Includes Louis Harris and Associates, Inc., survey on public attitudes about genetics.

10. Ibid.

11. Ibid.

12. Ibid.

13. Ibid.

14. Ibid.

15. National Science Foundation 1990 survey (n. 7).

16. Ibid.

17. OTA 1987 survey (n. 9).

18. Jay A. Winsten, "Science and the Media: The Boundaries of Truth," *Health Affairs* 4 (1985): 5–23.

19. Sharon M. Friedman, Sharon Dunwoody, and Carol L. Rogers, eds., *Scientists and Journalists: Reporting Science as News* (Washington, D.C.: American Association for the Advancement of Science, 1986).

Genetic Discrimination: The Use of Genetically Based Diagnostic and Prognostic Tests by Employers and Insurers

Larry Gostin

The Human Genome Project (HGP) is a worldwide research effort with the goal of mapping and sequencing an estimated one hundred thousand human genes.[1] The National Center for Human Genome Research anticipates that genetic information will be the source book for biomedical science in the twenty-first century.[2] Professor Robert Cook-Deegan, an ethicist at the center, explains that while science now conceptualizes disease in terms of biochemistry, it is possible to envision that science will move to the next stratum of reductionism, identifying and manipulating the genes that provide instructions for the biochemistry in every human cell.[3]

The HGP will enhance the ability of science and medicine to gather and organize information to foresee a person's future potential and disabilities. Enormous human benefits may ensue in understanding the etiology and pathophysiology of genetic disorders, preventing disease through genetic counseling, and treating the disorders through genetic manipulation.[4] Genomic information will help clinicians to understand and eventually to treat many of the more than four thousand known genetic diseases as well as those multifactorial diseases to which there is a genetic predisposition.[5]

The HGP, however, also has the potential for social detriment. Imagine that employers and insurers will one day be able to obtain a genetic profile from the blood drawn from a small finger prick.[6] The genetic profile will go well beyond the discrete conditions currently understood, such as Huntington's, sickle cell, cystic fibrosis, and Duchenne muscular dystrophy. The employer may delve into future probabilities for a wide range of physical conditions such as cancer, heart disease, Alzheimer's disease, and schizophrenia. Professor Lance Liebman observes that "suddenly the job applicant is not a member of an undifferentiated population . . . [but has] a statistically analyzable medical future."[7]

This essay analyzes the law, ethics, and public policy concerning "genetic discrimination," which is defined as the denial of rights, privileges, or opportunities on the basis of information obtained from genetically based diagnostic and prognostic tests.[8]

Prejudice, alienation, and exclusion often accompany genetic-associated diseases even though, by definition, the person has no control over the disorder and it is not the result of willful behavior. The fact that genetic diseases are sometimes closely associated with historically insular ethnic or racial groups such as African blacks,[9] Ashkenazi Jews,[10] or Armenians[11] only compounds the potential for invidious discrimination.[12] Preventing discrimination based on a person's health care status and providing equal opportunities for persons with disease and disability both emerge as powerful societal goals fully recognized by Congress[13] and the courts.[14]

Genetic discrimination, however, cannot always be attributed to pernicious myths, irrational fears, or ethnic hatred. The demand for genomic information can be rational. The HGP places squarely before society a set of legitimate values that conflict with the antidiscrimination principle.[15] These societal values include promoting the health of people known to have genetic predispositions to disease by excluding them from hazardous environments; ensuring that workers are fit for employment by requiring them to meet qualification standards; and saving costs by refusing to hire persons who have a high probability of disproportionately burdening company benefit plans. In an age when large employers pay as much, or more, for health and welfare benefits as for the raw materials of production, they are anxious to improve productivity and to lower personnel costs.[16] The HGP, thus, produces unequaled societal conflict between the rights of individuals and the rights of institutions. The goal of the law and ethics explored in this essay is to ensure that genomic information is used only when such use is clearly necessary to protect a person's health or safety or to enforce legitimate performance criteria for a job, service, or benefit.

The essay begins with a review of genetic discrimination, its prevalence and potential, as well as arguments as to why discrimination violates fundamental human rights principles and undermines the public health goals necessary to fulfill the true promise of the HGP. I then examine whether the Americans with Disabilities Act (ADA) and the corpus of antidiscrimination law are sufficiently relevant and comprehensive to safeguard against genetic discrimination. As part of this examination, the essay will discuss genetic discrimination in two key sectors: employment and insurance. I end with proposals for future legislative and judicial safeguards against genetic discrimination.

PREVALENCE OF AND POTENTIAL FOR GENETIC DISCRIMINATION: JUSTICE AND PUBLIC HEALTH

Amidst the euphoria generated by the HGP, the National Institutes of Health[17] and the Congress [18] have expressed concern that genomic information may result in stigma and discrimination. Indeed, fear of the social impact of discovery of the human genome may be the most significant impediment to continued full funding of the project.[19] Fortunately, most funders appreciate the notion that fear of unknown medical advances does not justify stifling scientific progress, but that it does require careful ethical planning and legal safeguards.

The Social Impact

Genetic discrimination violates basic tenets of individual justice and is detrimental to the public health. Discrimination based on actual or perceived genetic characteristics denies an individual equal opportunity because of a status over which he or she has no control. Genetic discrimination is as unjust as status-based discrimination against other historically disfavored classes, discrimination based on race, gender, or disability. In each case, people are treated inequitably not because of their abilities but solely because of predetermined characteristics. The right to be treated equally and according to one's abilities in all the diverse aspects of human endeavor is a core social value.

Genetic discrimination is socially harmful not merely because it violates core social values, but also because it thwarts the creativity and productivity of human beings—perhaps more than the disability itself. When qualified individuals are excluded from education, employment, government service, or insurance, the marketplace is robbed of skills, energy, and imagination. Such exclusion promotes physical and economic dependency which drains social institutions rather than enriching them.

Genetic discrimination, most significantly, undercuts the fundamental purpose of the HGP, which is to promote public health. The infusion of unprecedented human and financial resources into the HGP is justified by the promise of clinical benefits in identifying, preventing, and effectively intervening in human disease. If fear of discrimination drives people away from genetic diagnosis and prognosis, renders them less willing to confide in physicians and genetic counselors, and makes them more concerned with loss of jobs or insurance than with care and treatment, the benefits cannot be fully achieved.

The public health impact of discrimination will become all the more clear as genetic technology advances. The public health justification for widespread collection and utilization of genomic data will increase as genetic diagnosis and prognosis become more accurate and less expensive and as gene therapies become standard medical practice. The ability of society to develop and implement ambitious genetic screening and intervention strategies will depend upon the adequacy of safeguards against discrimination and breaches of confidence.

Discrimination and Scientific Uncertainty

Complex and often pernicious mythologies emerge from the public ignorance of genetically based diagnostic and prognostic tests. The common belief is that genetic testing, because it is done by scientific standards, is always accurate and highly predictive and that it illuminates an inevitable predestination of future disability in the individual or her offspring. Of course, the facts are diametrically opposite to the common belief. The precision of genetic-based diagnosis and prognosis is uncertain for many reasons. The sensitivity of genetic testing is limited by the known mutations in a target population. For example, screening can currently detect only 75 percent of cystic fibrosis (CF) chromosomes in the U.S. white population;[20] only 56 percent of at-risk couples will both be identified as carriers.[21] Professors Wilford and Fost[22] calculate that one out of every two couples from the general population identified to be at risk by CF population screening would be falsely labeled, with the potential for increased anxiety, discrimination,[23] and alteration of reproductive plans.

Predicting the nature, severity, and course of disease on the basis of a genetic marker is also fraught with difficulty. The date of onset of disease, the severity of symptoms, and the efficacy of treatment and management are highly variable with most genetic diseases.[24] Some people remain virtually symptom-free, while others progress to seriously disabling illness. A marker for Huntington chorea, for example, presents an aura of the inevitability of a relentless and progressive neurologic impairment but, in fact, a great deal of variability is apparent even within the same family.[25] In a disease such as neurofibromatosis, some will suffer marked disability of the nervous system, muscles, bones, and skin, while others will exhibit only minor pigmented spots on their bodies.[26]

Many genetic-associated diseases, unlike CF and Huntington's, are not attributed to a single gene mutation or genetic marker but are multifactorial. Their appearance can depend on a complex interaction of genetic and environmental factors that cannot be measured accurately. Current scientific assessments point to numerous poorly understood multifactorial diseases ranging from colon cancer, heart disease, and emphysema to schizophrenia, depression, and alcoholism.

Genetic-based diagnosis and prognosis, therefore, is characterized by marked heterogeneity: the reliability and predictive value of testing is limited by known mutations and prevalence in the target population; variability exists in the onset, presentation, and outcome of disease; and predictions are confounded by a multiplicity of genetic, biomedical, and environmental factors.

For all these reasons, significant scientific uncertainty surrounds much genetic testing. Genomic information may be highly beneficial for patients and health care professionals in making prevention, treatment, diet, lifestyle, or reproductive choices. Employers, insurers, educators, police, and others, however, will surely come to have access to genomic information. When genomic information is used by social institutions not to prevent or treat disease but to deny opportunity, exclude from work or benefits, remove health care coverage, or restrict liberty, a whole new dimension to the HGP becomes apparent. Adverse decisions in employment and insurance are particularly hurtful when they are rendered on the basis of false assumptions regarding the nature, accuracy, and predictive ability of genetic tests.

Genetic Screening and Monitoring

The U.S. Congress Office of Technology Assessment (OTA) provides the only systematic data about the past, current, and future use of genetic screening and monitoring in the workplace.[27] A comparable population of industrial companies, utilities, and trade unions was surveyed in 1982 and 1989 in order to provide trend data. The OTA surveys show that relatively few companies are currently utilizing genetic screening and monitoring. The 1989 survey found that a total of 20 health officers, from 330 of the Fortune 500 companies responding (6 percent), reported the use of genetic monitoring or screening at present (12) or in the past (8), compared with 18 (6 current and 12 past) in 1982. Although monitoring and screening are still not common, note that the incidence of companies conducting current screening was twice as great in 1989 as in 1982.

If there has been little or no real growth in the number of companies conducting genetic monitoring and screening in the workplace, what do companies foresee for the future? In 1982, fifty-five companies said they would possibly use genomic information in the next five years. The 1989 OTA survey reports that fewer companies anticipate using genetic testing and monitoring. However, in a survey of four hundred firms conducted in 1989 by Northwestern National Life Insurance Company, 15 percent of companies responded that by the year 2000 they planned to check the genetic status of prospective employees and their dependents before making an offer.[28]

The only available data from the insurance industry validates the findings of the OTA employment surveys. In 1989, the American Council on Life Insurance conducted the first study on whether insurers were beginning to perform genetic testing. The council reported that no insurance company was performing its own tests, but some did access known genomic information in their underwriting decisions.[29] The Health Insurance Association of America has formed its own committee and plans to survey its members in the use of genetic testing.[30]

The absence of expected growth in genetic testing and monitoring in industry and insurance should not provide grounds for complacency in anticipating the social impact. Employers and insurers claimed to choose genetic testing based upon its predictive value, scientific consensus, and cost.[31] The HGP will certainly promote rapid progress in human molecular genetics, which suggests increased occupational and insurance use in the future. Industry and insurers are not likely to routinely use genetic diagnosis that costs, say, two thousand to three thousand dollars per test. But as the technology becomes capable of identifying a battery of genetic conditions at a fraction of the current cost, the sheer competitive nature of industry and insurance may result in a vastly increased amount of testing. American industry is likely to turn to genetic diagnosis in the future for many of the same reasons that have driven the sharp increases in drug, polygraph, and general medical testing in the workplace.[32]

In the end, market forces may be the single greatest factor in generating increased use of genetic testing. Market researchers have projected U.S. sales of genetic tests to reach several hundred million dollars before the end of the decade.[33] The emergence of commercial interests in genetic-test development provides powerful incentives to lower costs of genetic testing and to place the technology within the reach of industry and insurance.

Once genetic testing gains a foothold within a particular industrial or insurance sector, the sheer competitive nature of the marketplace may require its wider adoption. Clearly, if some insurers or employers begin to make increasingly sophisticated genetic predictions of ill health and shortened life, the pressure on others to utilize the same technology will become irresistible.

If the marketplace itself is the only restraint on the proliferation of genetic technology, then once the technology comes down in price and demonstrates a cost-benefit advantage, widespread adoption of the technology will become inevitable. The need for legal safeguards against genetic discrimination in employment and insurance is apparent.

Genetic Discrimination

The 1982 OTA survey reported that, of the eighteen companies taking some action on the basis of genetic testing, seven had transferred or dismissed the "at-risk" employee.[34] The 1989 OTA survey reported very few instances of negative personnel decisions as a result of genetic monitoring or screening. Only two Fortune 500 companies reported ever rejecting a job applicant or transferring an employee primarily or partly based upon the results of genetic testing.[35]

It is certainly probable that self-reporting data by industry would not disclose the full extent of discrimination, especially since no systematic studies of genetic discrimination in employment and/or insurance have been undertaken. Professor Paul Billings and his colleagues did report twenty-nine cases of apparent genetic discrimination based upon replies to an advertisement.[36] Other anecdotal reports of genetic discrimination have appeared in the media.[37]

Reported cases of genetic discrimination reinforce the idea that adverse decisions are often based upon mythologies and misconceptions rather than real cost or safety concerns. Adverse decisions are taken because of the person's genetic status rather than actual disability, lack of qualification, or even accurate forecasting of future impairment.

A common misconception is that the presence of a genetic trait can be equated with actual disability without any need to demonstrate a cogent nexus between presence of the trait, current impairment, and inability to meet reasonable qualification standards. Discrimination affects heterozygotes (unaffected carriers), at-risk individuals (with a predisposition to disease), and persons who are asymptomatic or have a minor form of the disease, as well as those whose functioning is significantly impaired by the disease.

Several cases of discrimination were reported involving heterozygotes of sickle cell[38] or Gaucher's disease.[39] The genetic trait may affect future offspring but not the person herself. Discrimination against an unaffected carrier is particularly pernicious since the condition is irrelevant to the person's current or future health status or abilities. The discrimination is based upon the mythology that a heterozygote has the disease or will develop it.

"At-risk" persons have a propensity to develop genetic disease based upon a positive DNA test result or a family history.[40] Persons at risk are often treated as though they were currently impaired with the most severe form of the disease even though there is great variability in the onset and severity of symptoms. Persons with the marker for

Huntington's have been discriminated against because of this misconception: cases have involved rejection by adoption agencies, termination from a pilot's position, and rejection of life insurance coverage.[41]

Persons actually affected with genetic disease are often wrongly believed to have severe symptoms even if their impairments are relatively minor. At the most extreme, discrimination occurs against the "healthy ill," those who test positive for a hereditary condition but are asymptomatic.[42] In other cases, persons are discriminated against because of a genetic condition despite the fact that the symptoms do not interfere with performance. Persons with Charcot-Marie-Tooth Disease (CMT), a hereditary motor-sensory neuropathy, have been rejected from life, accident, or automobile insurance.[43] They were apparently very mildly affected with only a slight foot deformity. One case of discrimination occurred despite a favorable letter from the person's personal physician and the absence of any record of past ill health or accidents.[44]

The desire to save cost is implicit in many cases of genetic discrimination. A decision to deny health or life insurance because of a diagnosis of hemochromatosis[45] or a positive fetal DNA test for cystic fibrosis[46] is based on the belief that these individuals or their offspring could burden health care and other benefit plans. Whether cost alone is ever a sufficient justification for genetic discrimination becomes an overriding public policy question.

LEGAL MECHANISMS TO REDRESS GENETIC DISCRIMINATION

While discrimination against human beings because of their status may be morally wrong, it is not necessarily unlawful. Congress and state legislatures have proscribed discrimination against certain classes based on their inherent characteristics. These characteristics include race, gender, religion, national origin, age, and disability.[47] Equivalent statutory safeguards do not exist in most jurisdictions to protect against inequitable treatment based on other immutable personal characteristics such as sexual orientation, intelligence, height, looks, or personality. (These characteristics, however, could rise to the level of disabilities if they resulted in substantial impairments in life activities, as they might in persons with mental retardation or learning

disabilities.) The distinction among these diverse human characteristics is often hard to fathom. Yet the key issue in determining the
lawfulness of genetic discrimination is whether it is more akin to race,
gender, or disability than to sexual orientation, baldness, or personality. Certainly genetic discrimination which disproportionately burdens particular races, religions, or gender needs to be examined under
relevant civil-rights legislation. But before going into such obviously
fertile areas of inquiry, I will ask the more basic question: Can a genetic condition or trait be regarded as a disability?

The Applicability of Disability Law
to Genetic Discrimination

Three sources of disability law emerge from federal, state, and municipal legislatures: the Americans with Disabilities Act of 1990 and
other federal disability legislation,[48] state and municipal handicap laws,
and genetic-specific laws. The Americans with Disabilities Act of 1990
(ADA) is the most sweeping civil-rights measure since the Civil Rights
Act of 1964. The ADA, unlike its predecessor, the Rehabilitation Act of
1973,[49] extends antidiscrimination protection for persons with disabilities to the purely private sector in employment (Title I), public services
(Title II), public accommodations (Title III), and telecommunications
(Title IV).

State and municipal disability legislation stand as important supple-
ments to federal law. All states have disability statutes, all but four of
which prohibit discrimination in the private as well as the public sector.[50] Courts have reasoned that state disability statutes closely follow
the federal civil-rights approach and should be construed accord-
ingly.[51] State and local disability laws are characteristically enforced by
a well-established array of experienced state human-rights agencies
which can be far more productive than courts in preventing and reme-
dying discrimination. These administrative agencies initiate targeted
education and use negotiation as an effective tool.[52] Many agencies
report that they settle 80 percent or more of their cases through these
less-costly and less-adversarial methods.[53] State and local disability
legislation will continue to be important sources of law to prevent and
remedy genetic discrimination because of the existing regulatory network dedicated to education, fact-finding, and alternative forms of dispute resolution. The federal offices for civil rights (HHS and EO) also

employ negotiators to settle complaints; the ADA (ss13) encourages all types of alternative forms of dispute resolution.

To reinforce the importance of safeguarding persons from genetic discrimination, a few states and municipalities have enacted specific legislation. Some of these laws apply generally to hereditary disorders,[54] but most are directed to particular traits such as sickle cell, Tay-Sachs, or Cooley's anemia.[55] Most state and local legislatures have thus far refrained from enacting genetic-specific antidiscrimination legislation,[56] so the primary source of law continues to be disability law.

A thorough search of the legislative history of the ADA reveals that little attention was given to genetic discrimination. Congressman Steny Hoyer, the floor manager in the House, informed the Congressional Biomedical Ethics Advisory Committee that genetic discrimination was "not raised or discussed" and that it could not therefore be addressed by the conference committees. Congressman Hoyer recognized that genetic discrimination was "improper" and "very dangerous," but left it to the courts to determine whether it was covered under the ADA.[57]

Some of the legislative history of the ADA did, in fact, reflect the view that genetic discrimination was covered. Several members of Congress expressed the position that "the ADA will also benefit individuals who are identified through genetic tests as being carriers of a disease associated gene."[58] These legislators referred to the history of genetic discrimination, particularly during the sickle cell screening programs of the 1970s.[59]

Defining a disability. Disability is defined broadly in the ADA to mean: "a physical or mental impairment that substantially limits one or more of the major life activities,[60] . . . a record of such impairment, or . . . being regarded as having such an impairment."[61] "Physical or mental impairment" includes the following: any physiological disorder or condition, disfigurement, or anatomical loss affecting any of the major bodily systems, or any mental or psychological disorder such as mental retardation, mental illness, or dementia.

A person is disabled if he or she has a "record" of or is "regarded" as being disabled or is perceived to be disabled, even if there is no actual incapacity.[62] A "record" indicates that the person has a history of impairment, or has been misclassified as having an impairment. This provision is designed to protect persons who have recovered from a disability or disease which previously impaired their life activities.[63] By including those who have a record of impairment, Congress acknowledged that people who have recovered from diseases (such as cancer) or have diseases under control (such as diabetes) face discrimination based upon prejudice and irrational fear.[64]

The term "regarded" as being impaired includes individuals who do not have limitations on their major life functions, but are treated as if they did. This concept protects people who are discriminated against in the false belief that they are disabled. This provision is particularly important for individuals who are perceived to have stigmatic conditions that are viewed negatively by society. It is the reaction of society, rather than the disability itself, which deprives the person of equal enjoyment of rights and services.

Current genetic disability. Persons who are currently disabled by a genetic disease are undoubtedly covered under the ADA. The legislative history of the act as well as the prior case law make clear that disability is defined according to the degree of impairment of life functions and not the etiology. No distinction can be drawn between genetic and other causes of disabilities. The Congress and the courts have recognized disabilities of genetic origin (e.g., Down syndrome,[65] Duchenne muscular dystrophy,[66] and cystic fibrosis)[67] and multifactorial origin (e.g., congenital heart disease, schizophrenia, epilepsy, diabetes mellitus, and arthritis).[68] The question for courts is whether the person is currently disabled. It does not matter how she came to be disabled.

The courts require a "substantial" limitation of one or more major life activities. A genetic condition which does not cause substantial impairment may not constitute a disability.[69] If a person with neurofibromatosis, for example, has only mild changes in pigmentation she may not be disabled, but if she suffers from gross disfigurement she most assuredly would be protected under the ADA.[70] Citing the example of cosmetic disfigurement, the Supreme Court said that Congress was just as concerned about the effects of impairment on others as it was about its effects on the individuals.[71]

Future or predicted disability (presymptomatic). Genetic diagnosis or prognosis creates a new category of individuals who are asymptomatic but who are predicted to develop disease in the future. These individuals are sometimes referred to as the "healthy ill" or "at risk."[72] The number of people in this category is small, since the technology is new and has been applied only to relatively rare diseases such as

Huntington chorea. Progress in DNA technology will undoubtedly increase the range of disease which can be detected before symptoms appear.

Can a person who is currently healthy but who is predicted to become ill be classified as "disabled" within the meaning of the ADA? Strong reasons in law, ethics, and public policy suggest that the person should receive the same protection as those who are currently disabled. The ADA expressly protects not only individuals who are actually disabled but also those who are "regarded" or perceived to be disabled. The law, therefore, does not objectively measure the actual abilities or disabilities of the person. Rather, it judges the discriminator through his own subjective perceptions, prejudices, and stereotypes. Those who discriminate because of a future prediction of impairment are fostering a harmful stereotype because the person is currently healthy and capable of meeting all the performance criteria for the job, benefit, or service. The discrimination is speculative and prejudicial: no sound judgment can be made regarding whether, when, and to what extent the person will lose her skills and capabilities and whether reasonable accommodations could be provided. It would be inequitable for a defendant who intended to discriminate on the basis of disability to successfully raise the defense that the person was not, in fact, currently disabled. A narrow construction of the term "regarded" to be disabled would yield an anomalous result. Persons with genetic conditions would be required to wait until they actually developed symptoms before becoming eligible for protection.

The intention of Congress to include future disability is reflected both in the legislative history of the ADA and in the prior case law. During the conference report debate on the ADA, Congressman Hawkins opined that persons who are genetically at risk "may not be discriminated against simply because they may not be qualified for a job sometime in the future. The determination as to whether an individual is qualified must take place at the time of the employment decision, and may not be based on speculation regarding the future."[73] Several members of Congress concurred in this reasoning, expressly stating that future disability was covered by the term "regarded" as being disabled.[74]

The case law on the federal Rehabilitation Act and state disability statutes also suggests that it is unlawful to discriminate on the basis of future disability.[75] One court observed that it would be "ironic and insidious" if current disabilities were protected but the same protection were denied to those whom employers perceived to be predisposed to future disability.[76] The New York Court of Appeals held that obesity is a handicap even though it causes no current disability. The court aptly observed that "disabilities, particularly resulting from disease, often develop gradually. . . . An employer cannot deny employment simply because the condition has been detected before it has actually begun to produce deleterious effects."[77] Also, an employer may not point to future safety risks as grounds for dismissal.[78]

The analogy of asymptomatic HIV infection is helpful in ascertaining the court's likely position on persons at risk for genetic disease. A positive test for HIV infection is a powerful predictor of future disease and disability. In 1986 the Justice Department concluded that while the disabling effects of AIDS may constitute a handicap, pure asymptomatic infection could not.[79] The Justice Department position has been thoroughly repudiated by the Congress and the courts.[80] Both the legislative[81] and judicial[82] branches have made clear that pure asymptomatic infection is protected under disability law.

It is difficult to distinguish a predictive test for AIDS and one for Huntington's. In both cases there is no current disability, the predictive value of the test is strong, and the date of onset and severity of symptoms is uncertain. Employers might argue that persons who test positive for HIV have a current infection manifesting a clear disease process, while persons testing positive for Huntington's do not. This is not a convincing argument. A defect in a specific chromosome can be identified as the beginning of a genetic disease process in the same way as infection is identified as the beginning of a contagious disease process.

Public policy would clearly be skewed if it left individuals unprotected while free of symptoms and protected them only after they developed symptoms.[83] Fortunately, courts are not likely to accept this construction of the ADA.

Genetic carriers. Carriers of recessive genetic diseases such as hemoglobin disorders (e.g., sickle cell and thalassemia) and Tay-Sachs have only one gene that influences the disease, whereas two genes are required in order to manifest the disease. A carrier will not develop symptoms, but her offspring may inherit the disease if her partner is also a carrier. Employers, insurers, or health care providers may discriminate against carriers because of the fundamental misconception that a recessive gene might affect a person's own health or capabilities.[84]

The ADA's prohibition of discrimination based upon a perception of disability applies to those who falsely assume that carriers are, or will become, disabled. The primary purpose of disability law is to overcome such irrational fears and beliefs. The courts have made clear, for example, that an unfounded fear of contracting an infectious disease is not tolerated under the law.[85] A clear consensus emerges from the case law that employment decisions must be based upon reasonable medical judgments showing that the person's disability prevents her from meeting legitimate performance criteria.[86]

Multifactorial diseases and environmental factors. The social impact of genetic diagnosis and prognosis is felt strongly in the occupational setting. The identification of persons who are hypersusceptible to occupational disease provides an ostensible health justification for employment discrimination. These persons are protected under the ADA provided the employer's decision is based upon a perception that the person is, or will become, disabled. A federal district court, for example, held that a person who is unusually sensitive to tobacco smoke is handicapped under the federal Rehabilitation Act of 1973. The court reasoned that this hypersensitivity limited one of the employee's major life activities—i.e., his capacity to work in an environment which is not completely smoke-free.[87] Similarly, an athlete may be disabled if she has a weak heart[88] or poor eyesight,[89] both of which make her susceptible to future harm or injury. A firefighter may be disabled if the stresses of fire fighting would provoke a sickle cell crisis.[90]

The promise of the ADA is to protect all individuals who are, or are perceived to be, disabled in the past, present, or future. It would be a betrayal of that promise if the law did not equally protect individuals who are not and indeed may never become disabled, but whose predictive genetic tests cast a shadow over their own future health or that of their children.[91]

Employment Discrimination under Disability Law

"Qualification standards," including "direct threat." The antidiscrimination principle in the ADA applies only to "qualified individuals."[92] A "qualified" person must be capable of meeting all of the performance or eligibility criteria for the particular position, service, or benefit.[93] Qualification standards under the ADA may include a requirement that an individual shall not pose a "direct threat" to health

or safety in the workplace.[94] There is, moreover, an affirmative obligation to provide "reasonable accommodations"[95] or "reasonable modifications"[96] if they would enable the person to meet the performance or eligibility criteria. Employers are not required to provide a reasonable accommodation if it would pose an undue hardship on the operation of the business.[97]

How do such terms as "qualified," "direct threat," "reasonable accommodation," and "undue hardship" apply to discrimination based upon a genetic prediction of future disability for the employee or her offspring? Qualification standards are measured against current skills and performance. The fact that a person may become unqualified sometime in the future does not provide a justification for discrimination. As noted throughout the legislative process on the ADA, "the determination as to whether a person is qualified must be made at the time of the particular employment decision—of hiring, firing, or promotion—and may not be based upon speculation and predictions regarding the person's ability to be qualified for the job in the future."[98]

Congress, however, was acutely aware of potential risks to health and safety posed by disabled persons to employees or others in the workplace.[99] Although the direct-threat criterion was limited to persons with contagious disease in the Senate bill, it was extended in conference to all individuals with disabilities.[100] The ADA does not override any legitimate medical standards or requirements for workplace safety established by federal, state, or local law or by employers.[101]

Since occupational health and safety concerns are incorporated within qualification standards, they must primarily be determined by actual workplace risks and not speculation about risks which are theoretical, remote, or distant.[102] The courts, however, are likely to uphold decisions to protect employers from foreseeable risks in the immediate future, particularly for safety- or security-sensitive positions.[103] For example, clear medical evidence demonstrating a likelihood of harm or injury may disqualify a person from an inherently risky position such as airline pilot, police officer, or firefighter.[104]

The explicit language of the ADA refers to *significant* risks to the health or safety of *other* individuals in the *workplace*.[105] The language helps define the parameters of genetic discrimination based upon current or foreseeable health and safety risks. Occupational risks must be significant. The determination of significant risk must be based upon scientific evidence.[106] Disability law has been thoughtfully crafted to replace reflexive actions based upon irrational fears, speculation, stereotypes, or pernicious mythologies with carefully reasoned judgments based upon well-established scientific information.[107] Significant risk must be determined on a case-by-case basis, and not under any type of blanket rule, generalization about a class of disabled persons, or assumptions about the nature of disease. This requires a fact-specific individualized inquiry resulting in a "well-informed judgment grounded in a careful and open-minded weighing of risks and alternatives."[108] A specific determination must be made that a person with a genetic predisposition will develop symptomology in the immediate future leading to a real health or safety threat in the workplace.

The language of the ADA refers to risk to "other individuals" in the "workplace." Courts might disregard risks to the health of the disabled person on a strict reading of the act. Thus, it is conceivable that disabled persons may argue that they cannot be discriminated against merely because they are hypersusceptible to workplace toxins or because they pose a safety risk to themselves. Courts, however, may well take a broader view of the act, and disqualify disabled persons if occupational exposure poses a significant and immediate threat to their health. Courts are not likely to overturn employers' decisions to terminate employees to avoid an immediate sickle cell crisis[109] or to avoid a likely heart attack or a physical injury to a vulnerable athlete.[110] The EEOC regulations for the ADA, however, expand this to include risks to the worker's own health, but some commentators question whether courts will find these regulations valid.[111] Whether or not a well-documented and serious longer-term risk of cancer would justify discrimination remains to be decided.[112] The courts, however, would comply with the letter and spirit of the ADA if they did not allow discrimination based upon cumulative exposure to workplace toxins over many years. This could open the door to potentially widespread genetic discrimination. As suggested earlier, it would also provide an excuse for employers to weed out hypersusceptible workers rather than reducing overall toxic levels.

The reference in the ADA to the risk to others in the "workplace" raises the question of whether employers can discriminate based upon risks to future offspring. The language would not permit discrimination against genetic carriers since the risk to the fetus does not arise from the workplace environment. Risks of teratogenicity to the employee or congenital risks to the fetus in the womb stemming from occupational exposure may present closer jurisprudential and ethical questions. This form of discrimination is more likely to be directed against women rather than disabled persons, so it will be discussed under gender discrimination below.

"Reasonable accommodations" and "undue hardship." The employer has an affirmative obligation to provide reasonable accommodations to enable a person with a disability to meet qualification standards, including the reduction of significant risks to health or safety.[113] While Congress was primarily addressing physical barriers to access, it also required that workplace environments be "usable" by individuals with disabilities.[114] This could include reducing the environmental hazards to which the disabled person is hypersusceptible. One federal district court assumed that this would include marked reductions in tobacco smoke to accommodate an employee who was hypersensitive, but did not require providing a smoke-free environment.[115]

The ADA requires reasonable accommodations unless they would impose "undue hardship" on the operation of the business.[116] Undue hardship means an action requiring significant difficulty or expense when considered against such factors as the nature and cost of the accommodation and the overall financial resources and size of the business.[117] Employers do not have to provide accommodations that would fundamentally alter the nature of the industry;[118] a company need not stop producing batteries to eliminate lead levels. Employers that can improve the workplace environment without undue financial burden may be required to do so.

Preemployment genetic testing and prognosis. The ADA specifies that the prohibition of discrimination against persons with disabilities applies to medical examinations and inquiries.[119] Historically, employers gathered information concerning the applicant's physical and mental condition through application forms, interviews, and medical examinations. This information was often used to exclude persons with disabilities from employment—particularly applicants with hidden disabilities such as epilepsy, emotional illness, cancer, or HIV infection.[120] Employers' abilities systematically to obtain and use medical information to predict currently hidden conditions is vastly expanding with the development of genetic testing and prognosis.

Discrimination against persons with hidden disabilities has been technically unlawful since the Rehabilitation Act of 1973. Enforcement, however, has been exceedingly difficult, since an employer did not have to disclose that the person's medical condition was the prime reason for the failure to hire. So long as employers were able to conduct extensive medical examinations before offering a job, they could effectively hide the true reason for the employment decision.

The ADA's most radical departure from the Rehabilitation Act is its express proscription against pre-offer medical inquiries.[121] Section 102(c)(2) prohibits employers from conducting medical examinations or inquiries as to whether a job applicant is disabled. Preemployment inquiries must be limited to assessing the applicant's ability to perform job-related functions.[122] Thus, employers may not require job applicants to undergo extensive medical examinations and screenings, including diagnosis and prognosis for genetic traits or conditions such as Huntington's, sickle cell, or Tay-Sachs. This will strictly limit the employer's ability to obtain information about a person's current and future illness, diseases, or genetic predispositions before a job is offered.

The ADA permits an employer to require a medical examination only after an offer of employment has been made. All entering employees must be subjected to the same examination and the medical information must be kept strictly confidential.[123] Employers are also limited in their rights to conduct medical examinations or inquiries after a person is hired. The employer cannot compel an employee to take a medical examination or inquire as to whether the employee is disabled unless the examination or inquiry is job-related and consistent with business necessity.[124]

Congress, in enacting the ADA, recognized that a medical examination or inquiry that is not job-related serves no legitimate employment purpose but simply stigmatizes persons with disabilities.[125] The ADA will significantly impede the growing use of medical testing and information gathering used by employers across America, thus transforming the way the business community makes hiring decisions.

Employer costs: health, disability, life, and other insurance benefits.[126] When asked, employers cite cost-benefit and greater productivity as the main reasons for medical testing.[127] Costs to employers for health care and other benefits have risen substantially in recent years.[128] These costs are borne directly by employers with larger work forces, many of which tend to be self-insurers or have experience rating. Alternatively, increased benefits costs are passed on from insurance companies to employers.

Some astute commentators on disability law give considerable credence to employer decisions not to hire persons with current or future disabilities based upon probable costs to health, disability, life, and other insurance benefits. "If a worker will become ill, and if the employer will be responsible for the medical costs as well as the output costs of the worker's absence, then the predicted illness is nothing but a future dollar cost that the employer must consider and discount."[129]

While industry certainly has made its fiscal position clear,[130] a legislative and judicial consensus has emerged that financial burdens on employee benefit programs do not justify discrimination. The Senate Labor and Human Resources Committee echoed a theme found throughout the legislative process: "An employer may not refuse to hire an individual because of fears regarding increased insurance costs attendant on hiring the individual—either because of increased costs to be incurred because of that individual's health needs or because of the health needs of that individual's family."[131]

Disability law is concerned only with the relationship between a person's disability and her ability to perform the job she seeks. The fact that the person may have an undesirable impact on disability, life, or health insurance programs is irrelevant.[132] The courts have also rejected a wide array of business and economic interests as justifications for discrimination: picketing the establishment,[133] a threat of violence against workers,[134] adverse publicity,[135] and loss of clientele.[136] The reasons that courts reject cost as an excuse for discrimination are that antidiscrimination is considered a higher social value than strict business interests; and that cost incentives, if allowed, would swallow up all the protection currently conferred by disability law.[137]

People with genetic traits, conditions, or predispositions must not be denied jobs or promotions based upon their perceived or future disabilities, and they must have equal access to health and other insurance coverage provided to all employees. Employers, however, may circumvent antidiscrimination principles when they are acting as self-insurers, as the following section demonstrates. Agencies required to promulgate regulations under the ADA should carefully consider the interaction of unlawful employment decisions with lawful underwriting decisions to ensure that the latter are not used to circumvent the overriding antidiscrimination principles under the act.

Insurance Discrimination under Disability Law

Congress intended to afford to insurers, employers, and health care providers the same opportunities they would enjoy in the absence of the ADA to design and administer insurance products and benefit plans in a manner that is consistent with the basic principles of underwriting, classifying, and administering risks.[138] Thus, insurers may continue to sell to and underwrite individuals applying for life, health, or other insurance;[139] and employers and their agents may establish and observe the terms of employee benefit plans based upon sound actuarial data.[140]

The ADA, therefore, does not restrict an insurer, health care provider, or any entity that administers benefit plans from carrying on its normal underwriting activities. This includes the use of preexisting condition clauses in health insurance contracts, the placing of caps or other limits on coverage for certain procedures or treatments, and the charging of premiums to persons with higher risks.[141]

The adverse impact of underwriting for persons with genetic predispositions is significant. If insurers have actuarial data demonstrating a likelihood of future illness, they can limit coverage. More worrisome would be a decision by an insurer to view a genetic predisposition as a preexisting condition. The greater the predictive value of the genetic test, the more likely it is that insurers will regard the condition as uninsurable or preexisting. This process became clear in the reaction of insurance companies to HIV infection. As epidemiologic evidence demonstrated the inexorable course of HIV infection, insurers began insisting on having access to serologic information, and now much of the industry conducts its own HIV testing.[142] HIV infection is almost universally regarded as an uninsurable condition. As genetic tests begin to demonstrate the inevitability of future illness, we can expect insurers to follow the same course as has occurred with HIV infection.

The major limitation of the ADA on insurers, including self-insurers, is to prevent them from using underwriting as a subterfuge for invidious discrimination.[143] Congress intended a liberal construction of the word "subterfuge," so that any evasion of the principles of antidiscrimination—whether malicious, purposeful, or inadvertent—is unlawful.[144] Thus, insurers could not completely deny health insurance coverage on the basis of a genetic predisposition; and employers could not deny a qualified applicant a job because the employer's insurance plan did not cover the genetic disability or because of the increased cost.[145] The sharp distinction drawn by the ADA is that any discrimination among disabled applicants for insurance or employment must be justified on the basis of actuarial data demonstrating a heightened risk of future illness.

Whether one regards the ADA's exemption of underwriting as reasonable or not depends upon how the insurance industry is viewed. If the industry is regarded strictly as a business, it is difficult to challenge its right to discriminate on the basis of sound actuarial data. The very essence of underwriting is to classify people according to risk and to treat those with higher risks differently. Seen through the eyes of a business, no rational distinction could be drawn between genetic prognosis and smoking, hypertension, high serum cholesterol, or HIV infection. In each case, the health data can provide powerful predictions of future health and longevity. Worse still from the insurer's perspective is "adverse selection," where the buyer has access to genomic information which is denied to the company. Perhaps the greatest fear of insurers is that genetic testing will become common in clinical medicine and they will be barred from obtaining that information.

If the health insurance industry is viewed as an instrument of social policy, however, then the ADA's exemption of underwriting becomes troubling. The social purpose of insurance is to spread risk across groups, enabling wider access to health care services. If health insurance becomes unavailable or unaffordable to those who are most likely to become ill, then the social purpose of insurance is thwarted.

Once the pool of applicants is no longer an undifferentiated mass, and each person's medical future can be predicted with specificity, the conceptual foundations of the industry are turned around. It may not be too far in the future before a person's genome becomes a readable template for a wide array of diseases. If insurers choose to utilize genomic information to make sound actuarial predictions of disease, the ADA will, in all probability, be ineffective in placing any restrictions on the industry.

Genetic Discrimination: Disparate Impact on Race, Ethnicity, or Gender

Genetic prognosis, by its very nature, often has disproportionate impact upon vulnerable classes based on race, ethnicity, national origin, or gender. Sickle cell is associated with persons of African heritage, Tay-Sachs and Gaucher's disease (adult form) with Ashkenazi Jews, and familial Mediterranean fever with Armenians. Risks of teratogenicity or congenital risks to the fetus are often focused on women or pregnant women, setting up a class based on gender.

Title VII of the Civil Rights Act of 1964 [146] prohibits job discrimination based upon race or gender unless the discriminatory practice is related to job performance. Lack of discriminatory purpose is irrelevant.[147] Sandra Day O'Connor explained that the intention behind Title VII was "to prohibit an employer from singling out an employee by race or sex for the purpose of imposing a greater burden or denying an equal benefit because of a characteristic statistically identifiable with the group but empirically false in many individual cases."[148]

The outcome of Title VII lawsuits is often dependent upon whether the racial or gender discrimination is intentional. If the class is explicitly based upon race or gender or if the discrimination is intentional, the employer is required to meet the more exacting standard that the action is based upon a "bona fide occupational qualification." Title VII also allows lawsuits based upon the disparate impact of an apparently neutral policy. If neither discriminating intent nor an explicit racial or gender bias is shown, but only a discriminatory effect, the employer need only demonstrate that there was "business necessity" for the employment practice.[149]

Surprisingly little court litigation has focused on the burden of genetic testing on racial minorities or women.[150] An immediate problem is that a genetic prognosis represents a racially neutral policy which does not expressly discriminate on the basis of race or gender.[151] It is likely that the great majority of Title VII suits brought to remedy genetic discrimination will be based upon disparate impact theory—viz., the genetic trait or condition disqualifies proportionately more racial minorities or women.

Sickle cell classification: a form of race discrimination? Some courts have said in dicta that a job classification based upon sickle cell anemia or trait would create a disparate impact on African Americans.[152] Yet no lawsuit has been successful under this theory. In *Smith v. Olin Chemical Corporation,* an African American employee brought a Title VII claim in federal court when he was fired due to current and future bone degeneration, which is characteristic of sickle cell anemia. The Fifth Circuit Court of Appeals rejected his claim that he was dismissed on account of his race, stating that the category of "bone degeneration" was racially neutral. The court also rejected arguments based upon disparate impact because of the "manifest job-relatedness of the requirement that a manual laborer have a good back."[153]

Employment decisions based upon the sickle cell *trait* may well violate Title VII. A class based upon sickle cell disproportionately impacts

African Americans. Since the existence of a genetic trait usually does not indicate any current or future illness, it is difficult to conceive how employers would justify the discrimination as a business necessity.

Fetal protection policies: a form of gender discrimination? An employer's decision not to hire women of childbearing age based upon the risk of teratogenicity or harms to the fetus in utero from occupational exposure may be easier to challenge because of a more direct gender classification and less obvious job-relatedness. Fetal protection policies, in one form or another, go back a long way. Earlier this century, the Supreme Court upheld exclusion of women from hazardous employment to protect the "future well-being of the race."[154] Since that time, however, a series of federal statutes set standards requiring employers to reduce toxic levels and other hazards to both men and women.[155] The Pregnancy Discrimination Act of 1978[156] amended Title VII to clarify that the statute prohibits "discrimination [against working women] on the basis of their childbearing capacity."[157]

The Supreme Court recently decided the question of whether fetal protection policies are allowed under Title VII. In *International Union, UAW v. Johnson Controls*,[158] the court held that sex-specific fetal-protection policies are unlawful. The case concerned a battery manufacturer's policy that barred any woman from working in a job that exceeded present lead levels, unless she presented medical evidence of sterility.[159] The Seventh Circuit found no violation of Title VII.[160] It reasoned that the fetal-protection policy constituted a gender-neutral rule which had a disparate impact on a protected group.[161] The policy was justified by the "business necessity" of preventing "substantial health risk" to the children of female, but not male, workers.[162]

The Supreme Court found that Johnson Controls' fetal-protection policy constituted facial discrimination, as it applied to female employees, but not to male employees.[163] Such a policy could be justified only if sex were shown to be a "bona fide occupational qualification."[164] The court noted that the BFOQ defense is a narrow one, and that legislative history and case law forbid discrimination against a woman because of her ability to become pregnant unless it interferes with her job performance.[165]

The court also addressed the issue of potential harm to fetuses, stating that "decisions about the welfare of future children must be left to the parents who conceive, bear, support, and raise them rather than to the employers who hire the parents."[166] Ultimately, of course, the resolution of this ethical and legal dilemma is to reduce environmental hazards that harm both men and women, rather than excluding a class of persons deemed hypersusceptible. Allowing employers to "fix the worker, not the job," [167] would harm the public health. The Occupational Safety and Health Administration (OSHA) already mandates that employers maintain a workplace "free from recognized hazards that are causing or are likely to cause death or serious physical harm," and it requires the secretary of labor to promulgate health and safety standards, to the extent feasible, such "that no employee will suffer material impairment of health or functional capacity." [168] Private employers should not be permitted to make determinations about what is "safe" for whole subclasses of the employed population.

The fetal-protection policy in *Johnson Controls* is directed more toward risks in utero than to genetic risks. Suppose an employer were to discriminate against women based upon the teratogenicity of workplace toxins. That case would raise squarely the issue of whether teratogenic risks fall exclusively, or even predominantly, on women. Medical evidence that toxic exposure causes deformed sperm leading to birth defects would surely result in a finding of intentional gender discrimination. In such circumstances, Title VII would not tolerate consigning women to a class of employment with lower pay and potential based strictly on the possibility they could become pregnant.

Genetic-Specific Antidiscrimination Statutes

A minority of states have enacted statutes which specifically apply to persons with hereditary conditions. [169] The most progressive of these statutes recognize that disease-specific legislation might prove too rigid as scientific understanding of the human genome progresses. [170] These statutes have broad application to "any hereditary disorder," [171] and they draw the distinction between carriers and those who experience manifestations of the disease. [172] Other statutes are directed more narrowly to specific conditions or traits such as sickle cell, [173] PKU, hemophilia, cystic fibrosis, Tay-Sachs, or Cooley's anemia. [174]

Genetic-specific statutes around the country do not appear to follow any coherent policy or pattern. Only a few genetic-specific statutes ban discrimination. California adopted a general antidiscrimination policy which includes penalties for violating the Hereditary Disorders Act of 1990. The California statute is comprehensive and prohibits

"stigmatization" and "discrimination" against "carriers of most deleterious genes." It also proscribes mandated state restrictions on childbearing decisions regardless of the genetic purpose.[175]

Statutes in Florida, Illinois, Louisiana, New Jersey, New York, and
North Carolina are patterned after disability law and prohibit employment discrimination against persons with any "atypical hereditary or
blood trait"[176] or particular genetic conditions or traits.[177] In some
cases the statutes have more general application beyond employment
discrimination, including disparate impact on racial or ethnic minorities[178] or discrimination on the basis of familial status when in the
process of securing legal custody.[179]

The remainder of the genetic-specific statutes prohibit certain types
of screening,[180] provide funding for research or treatment,[181] or require mandatory information on genetic disorders to be given to marriage applicants.[182] Others are concerned with genetic counseling and
confidentiality.[183]

A review of current state statutes reveals a patchwork of provisions
which are incomplete and inconsistent, and which fail to follow a
coherent vision for genetic screening, counseling, treatment, and
discrimination.

FUTURE LEGAL SAFEGUARDS
AGAINST GENETIC DISCRIMINATION

The course currently being charted by the HGP is filled with the
promise of unimagined medical advancement for humankind. Unfortunately, the potential to harm human beings by rendering them virtually unemployable or uninsurable is just as real.[184] Policymakers
should be considering several legal strategies to safeguard against
genetic discrimination. While the ADA emerges as a mighty tool to
remedy genetic discrimination, it would be preferable not to leave its
construction to the vagaries of future adjudication. The ADA specifically redresses discrimination based upon past disability ("record of
impairment"), current disability ("impairment"), or perception of disability ("regarded" as impaired). The ADA, however, is silent about
discrimination based upon future disability. While this essay has presented a number of legal, ethical, and public policy arguments to suggest that future disability is covered, a simple amendment to section 3

of the ADA would remove any uncertainty. A new subsection "D" would amend the definition of disability to include "having a genetic or other medically identified potential of, or predisposition toward, such an impairment." Such language would ensure that discrimination against currently asymptomatic persons based upon a future prediction of disease would be covered by the ADA.

There remains in the ADA one major gap in coverage which is not so easily rectified. That is the exclusion of insurance underwriting from the coverage of the act. Future regulations on the ADA should seek to ensure that employers do not use this underwriting exclusion to discriminate against applicants or employees based upon financial burden. Employers may well turn to imaginative actuarial justifications for invidious discrimination. Employers need not take the drastic measure of not employing a person. Rather, employers may make the benefits plan so unattractive and unaffordable as to erect a formidable barrier to disabled persons.

Strict regulations preventing employers from evading the principles of the ADA are only a stopgap remedy. Ultimately, a political choice will have to be made by Congress as to whether insurance and self-insurance are merely businesses, or whether they have a wider social purpose. If insurance discrimination rises to a truly unconscionable level as the genome is mapped, then society may have to confront the issue of access to health care.

Gaps in the coverage of disability law also need to be addressed at the state level. State disability laws should also be amended to make clear that they cover future disability. The continued importance of state disability law cannot be underestimated, given the much more efficient administrative structures and greater resources available to combat discrimination.

Policymakers will also have to confront the philosophic and pragmatic issue about whether genetic-specific legislation is necessary or desirable. The great majority of state legislatures have enacted HIV-specific legislation [185] so the precedent is set for the concept of separately addressing particularly vexing public health problems. The national HGP should give careful consideration to developing a set of legislative guidelines for future genetic-specific statutes. In the absence of such guidelines, the prospect for rational legislation appears low.

Model guidelines on the law and ethics of genetic screening, confidentiality, and discrimination ought to go hand in hand with scientific advances. Funding, original thinking, and carefully crafted policy on the legal and ethical dimensions of the HGP are as essential as the science itself.

NOTES

This essay was previously published in the *American Journal of Law and Medicine* 17 (1992): 109–144. The copyright for the article is held by the American Society of Law, Medicine, and Ethics, and the Boston University School of Law; the revised paper is reprinted with permission.

1. Two reports published in 1988 laid the groundwork for the HGP by setting out methods and goals. U.S. Congress, Office of Technology Assessment, *Mapping Our Genes: Genome Projects: How Big, How Fast?* (1988); National Research Council, *Mapping and Sequencing the Human Genome* (1988). A report from the U.S. Public Health Service and Department of Energy in 1990 updates earlier descriptions. *Understanding Our Genetic Inheritance: The U.S. Human Genome Project: The First Five Years FY 1991–1995* (1990) (hereinafter cited as *The Human Genome Project*). See also J. D. Watson and R. M. Cook-Deegan, "The Human Genome Project and International Health," *Journal of the American Medical Association* 263 (1990): 3322; G. J. Annas, "Mapping the Human Genome and the Meaning of Monster Mythology," *Emory L. J.* 3, no. 3 (Summer 1990).

2. *The Human Genome Project*, at vii.

3. R. M. Cook-Deegan, "Social and Ethical Implications of Advances in Human Genetics," *Southern Medical Journal* 83 (1990): 879.

4. See, e.g., S. E. Antonarakis, "The Mapping and Sequencing of the Human Genome," *Southern Medical Journal* 83 (1990): 876; J. L. Goldstein and M. S. Brown, "Genetic Aspects of Disease," in E. Braunwald, K. J. Isselbacher, R. G. Petersdorf et al., eds., *Harrison's Principles of Internal Medicine*, 11th ed. (1989), p. 285 (hereinafter cited as *Principles of Internal Medicine*).

5. *The Human Genome Project*, at viii.

6. L. Liebman, "Too Much Information: Predictions of Employee Disease and the Fringe Benefit System," *U. Chicago Legal Forum* (1988): 57, 60; Hening, "Body and Mind: High-Tech Fortunetelling," *N.Y. Times*, Dec. 24, 1989, sec. 6, at 20.

7. Liebman, "Too Much Information," p. 60. See also Jerry E. Bishop and Michael Waldholz, *Genome: The Story of Our Astonishing Attempt to Map All the Genes in the Human Body* (New York: Simon and Schuster, 1990).

8. A similar definition was offered by P. R. Billings, M. A. Kohn, M. de Cuevas, J. Beckwith, J. S. Alper, M. R. Natowicz, "Discrimination as a Consequence of Genetic Testing," *Am. J. Human Gen.* 50: 476–482 (hereinafter cited as Billings et al.) ("differential treatment based on apparent or perceived human variation presumed to have a genetic origin").

9. E.g., sickle cell disease.

10. E.g., Tay-Sachs disease, Bloom's syndrome, Gaucher's disease (adult form).

11. E.g., familial Mediterranean fever.

12. *Principles of Internal Medicine*, p. 288.

13. Americans with Disabilities Act of 1990, Pub. L. No. 101-336, 104 Stat. 329, sec. 2 (findings and purposes).

14. School Board of Nassau County, Fla. v. Arline, 107 S.Ct. 1123 (1987).

15. See the excellent review in Liebman, "Too Much Information," p. 57.

16. See generally, Rothstein, "Employee Selection Based upon Susceptibility to Occupational Illness," *Mich. L. Rev.* 81 (1983): 1379.

17. *The Human Genome Project*, appendix 7.

18. Departments of Labor, Health and Human Services, and Education, and Related Agencies Appropriation Act, 1991. 136 Cong. Rec. H4996, H5003, daily ed. July 19, 1990 (the human genome project may divide us into "two groups, those with pluperfect and imperfect genes. . . . Taxpayers should not be put in the position of financing government programs without protections to ensure that those programs will not in the end lead to fencing them out of jobs or reasonably priced health insurance." We need to develop legal and ethical safeguards "before the knowledge genie is completely out of the bottle").

19. *Id.*

20. Screening for CF is now very much on the public agenda, despite cautionary statements by the Public Health Service and professional organizations. Statement from the National Institutes of Health Workshop on Population Screening for the Cystic Fibrosis Gene, *N. Engl. J. Med.* 323 (1990): 70 (hereinafter cited as *NIH Consensus Statement*); "The American Society of Human Genetics Statement on Cystic Fibrosis Screening," *Am. J. Hum. Genet.* 46 (1990): 393.

21. *NIH Consensus Statement*, p. 70; B. Wilford and N. Fost, "The Cystic Fibrosis Gene: Medical and Social Implications for Heterozygote Detection," *JAMA* 263 (1990): 2777, 2779.

22. *Id.*, at 2779.

23. Professor Billings and his colleagues recount a case of a family with a child who has CF and received care through an HMO. When a second pregnancy occurred, a prenatal DNA test was positive for two copies of the CF gene. The HMO considered withdrawal of coverage if the family proceeded

with the pregnancy. Threats of legal action were required before the HMO agreed to continue coverage. Billings et al.

24. See generally, V. A. McKusick, "Mendelian Disorders," in A. M. Harvey, R. J. Johns, V. A. McKusick et al., eds., *The Principles and Practice of Medicine*, 22d ed. (1988), p. 281 (hereinafter cited as *Principles and Practice of Medicine*).

25. M. R. De Long and H. Moses, "Disorders of Movement," in *Principles and Practice of Medicine*, p. 1069.

26. McKusick, "Mendelian Disorders," p. 287.

27. Office of Technology Assessment, U.S. Congress, *The Role of Genetic Testing in the Prevention of Occupational Disease* (1983): 33–46 (hereinafter cited as *OTA 1983 Report*); Office of Technology Assessment, *Genetic Monitoring and Screening in the Workplace* (1990): 171–188 (hereinafter cited as *OTA 1990 Report*). Genetic testing includes a number of technologies to detect genetic traits, changes in chromosomes, or changes in DNA. The OTA distinguishes between two different kinds of genetic testing: examining persons for evidence of induced change in their genetic material (monitoring) and identifying individuals with particular inherited traits or disorders (screening). *Id.*, at 3–6. The OTA terminology is somewhat confusing from a public health perspective since testing usually refers to case identification of an individual, while screening involves more systematic application to whole populations. See generally L. Gostin, W. Curran, and F. Clark, "The Case against Compulsory Casefinding in Controlling AIDS: Testing, Screening and Reporting," *Am. J. Law & Med.* 12 (1987): 7.

28. Reported in N. Holtzman, S. Brownlee, and J. Silberner, "The Assurances of Genes," *U.S. News and World Report*, July 23, 1990, p. 57.

29. R. Pokorski, *The Potential Role of Genetic Testing in Risk Classification* (American Council of Life Insurance, 1990). See M. Sit, "Will Genetic Mapping Threaten Workers' Privacy?" *Boston Globe*, August 21, 1990, p. 23.

30. Health Insurance Association of America, *Working Group on Genetic Testing* (1990) (staffed by J. Payne).

31. *OTA 1983 Report*, p. 36. The OTA found, however, that none of the genetic tests evaluated at that time met established scientific criteria for routine use in an occupational setting. The chasm in perception between the OTA and industry appears to lie in the employer's willingness to assume that if tests are sufficiently reliable for clinical use, they can safely be used in occupational settings. See also, "Prediction and Prejudice: Forging a New Underclass," *Consumer Reports* 55 (July 1990): 483.

32. A comprehensive OTA report on testing in health insurance documents the substantial rise in testing, including prospects for genetic testing. Office of Technology Assessment, U.S. Congress, *Medical Testing and Health Insurance* (1988).

33. The commercial appeal of genetic testing is revealed in a staff background paper to the *OTA 1990 Report*. M. Hewitt and N. Holtzman, "The Commercial Development of Tests for Human Genetic Disorders" (February 1988; unpublished manuscript). This paper reviews predictions of the market value of genetic tests from the following sources: Robert S. First, Inc., *Genetic Testing in the USA 1986–1990* (1986) ($550 million by 1990); "Arthur D. Little Projects a $5.7 Billion Clinical Diagnostic Market in 1990," *Genetic Engineering News*, March 1987, at 13 ($300–$500 million by 1995); "Disease Disposition Screening," *Biomedical Bus. Int'l.* 230, 230–232 (1986) (U.S. current market value of genetic tests estimated at $210 million); "DNA Probes Nudge Monoclonals in the Race to Exploit the Medical Diagnostics Market," *Genetic Engineering News*, September 1986, at 1, 12, 13, 21 ($500 million market value by 1993); "Market for DNA Probe Tests for Genetic Diseases," *Genetic Tech. News*, November 1986, at 6–7 ($950–$1,000 million market value by 1992). More recently, *Business Week* predicted a $200-million-a-year market for genetic tests being actively sought after by prominent biotechnology companies. Carey, "The Genetic Age," *Business Week*, May 28, 1990, at 68.

34. *OTA 1983 Report*, p. 37.

35. *OTA 1990 Report*, p. 182.

36. Billings et al.

37. See, e.g., V. MacDonald, "Ethical Eye on Insurers' Genetic Tests," *Daily Telegraph*, July 15, 1990, p. 9.

38. See E. D. Shapiro, "Dangers of DNA: It Ain't Just Fingerprints," *N.Y.L.J.* 203, no. 15 (January 23, 1990): 1 (Dr. Bereano's testimony before a congressional subcommittee: "The Armed Forces for many years has followed a policy of excluding carriers of sickle cell disease, despite the fact that these individuals are not themselves impaired."); "Air Force Rejects Cadets with Sickle Cell Trait," *N.Y. Times*, Feb. 6, 1980, p. C-10; Matthewman, "Title VII and Genetic Testing: Can Your Genes Screen You Out of a Job?" *How. L. J.* 27 (1984): 1199; Raymann, "Sickle Cell Trait and the Aviator," *Av. Space & Envtl. Med.* 50 (1979): 1170.

39. Billings et al. (denied government job because he was a "carrier, like sickle cell").

40. One newspaper, for example, reports the case of a Chicago woman turned down for a job after her prospective employer learned that her mother was schizophrenic, based on the fear that it was an inherited trait. Shaw, "Genetic Gains Raise Fear of a New Kind of Bias," *Philadelphia Inquirer*, Nov. 23, 1990, 1-A, 10-A.

41. Billings et al.

42. Billings et al. report a case of a person with hereditary hemochromatosis who, despite the absence of symptoms, was consistently denied insurance. See also Holtzman, Brownlee, and Silberner, "The Assurances of Genes," p. 57.

43. See J. W. Griffith and D. R. Cornblath, "Peripheral Neuropathies," in *Principles and Practice of Medicine*, pp. 1095–1096 (weakness, particularly footdrop, foot deformity, and hand weakness are the most severe manifestations of CMT).

44. Billings et al.

45. MacDonald, "Ethical Eye on Insurers' Genetic Tests," p. 9. Hemochromatosis is characterized by an excessive absorption and storage of iron and can be controlled.

46. Billings et al.

47. See nn. 145–167 infra and accompanying text.

48. Federal legislation provides protection to persons with disabilities in several areas: (1) the Fair Housing Amendments of 1989 is the prime legislation protecting disabled persons from discrimination in housing; and (2) the Education for All Handicapped Children Act (EHCA) gives all school-aged children with disabilities the right to a free public education in the least restrictive environment appropriate to their needs. (The EHCA was renamed by Congress to follow the more progressive language of "persons with disabilities." It is now entitled the Individuals with Disabilities Education Act [IDEA].)

49. The Rehabilitation Act of 1973 endures beyond the ADA and continues to be the prime legislation affecting persons with disabilities working for the federal government. See ADA, sec. 509 (coverage of Congress and the agencies of the legislative branch).

50. National Gay Rights Advocates, *Protection against Discrimination under State Handicap Laws: A Fifty-State Analysis* (1989). See M. Rowe and B. Bridgham, *AIDS and Discrimination: A Review of State Laws that Affect HIV Infection: 1983–1988* (Washington, D.C.: AIDS Policy Center, Intergovernmental Health Policy Project, George Washington University, 1989).

51. See, e.g., Raytheon v. Fair Employment and Housing Commission, Estate of Chadbourne, 261 Cal. Rptr. 197 (Cal. App. 2d Dist. 1989).

52. The ADA also encourages the use of alternative means of dispute resolution including settlement negotiations, conciliation, facilitation, medication, fact-finding, minitrials, and arbitration (sec. 513).

53. R. Steele, S. Karsten, B. Lorenz, and J. Ritter, *Identification and Assessment of State and Local Strategies to Prevent Discrimination* (1989).

54. See, e.g., Cal. Health and Safety Code, sec. 150(f) (West 1990) ("carriers of most deleterious genes should not be stigmatized and should not be discriminated against by any person . . .").

55. See, e.g., Civil Rights Amendments of 1990 in relation to the employment of persons with certain genetic disorders, 1990 N.Y. Laws 900, 1990 N.Y. A.N. 9437.

56. Compare with the flood of HIV-specific antidiscrimination legislation. See Rowe and Bridgham, *supra* n. 50; L. Gostin, "The AIDS Litigation Proj-

ect: A National Review of Court and Human Rights Commission Decisions, Part II: Discrimination," *JAMA* 263 (April 18, 1990): 2086–2093; L. Gostin, "Public Health Strategies for Confronting AIDS: Legislative and Regulatory Policy in the United States," *JAMA* 261 (March 17, 1990): 1621–1630.

57. Letter from Congressman Steny H. Hoyer to Alexander M. Capron, chair of the Biomedical Ethics Advisory Committee, August 1, 1990.

58. Rep. Augustus Hawkins, Conference Report on S. 933, Americans with Disabilities Act of 1990, 136 Cong. Rec. H4614 (daily ed. July 12, 1990).

59. *Id.*, Rep. Don Edwards, H4624; Rep. Henry Waxman, H4626.

60. "The term physical or mental impairment does not include simple physical characteristics, such as blue eyes or black hair. . . . [Nor does it include] environmental, cultural, and economic disadvantages" in and of themselves. S. Rep. No. 116, 101st Cong., 1st Sess. (1989), at 22. The question may arise as to why genetic traits for sickle cell or cystic fibrosis ought to be covered in the ADA but not the genetic determinants for blue eyes or black hair. The reason is simply that Congress has designated disability, but not general personal characteristics, under civil rights. *Id.*

61. Pub. L. No. 101-336, sec. 3(2). The definition of disability in the ADA is comparable to the term "handicap" in section 7(8)(b) of the Rehabilitation Act of 1973 and section 802(h) of the Fair Housing Act. Congress intended that regulations implementing the Rehabilitation Act (42 Fed. Reg. 22685 *et seq.*, May 4, 1977) and the Fair Housing Amendments apply to the term "disability" in the ADA. The use of the term "disability" instead of "handicap" represents currently acceptable terminology. See "The Americans with Disabilities Act of 1989: Reports of the Committee on Labor and Human Resources," S. Rep. No. 116, *supra* n. 60, p. 21.

62. This concept derives from Southeastern Community College v. Davis, 442 U.S. 397 (1979).

63. S. Rep. (Labor and Human Resources Committee) No. 116, *supra* n. 60, Aug. 30, 1989 (to accompany S. 933), p. 23.

64. School Board of Nassau County, Fla. v. Arline, 480 U.S. 273, 284 (1987).

65. See Bowen v. American Hospital Ass'n, 476 U.S. 610 (1986). (While the case turned on whether withholding of treatment for an esophageal obstruction was covered under the Rehabilitation Act, Justice White clearly saw Down syndrome as a protected handicap.) The Senate Committee on Labor and Human Resources drew attention to a New Jersey zookeeper who refused to admit children with Down syndrome because he feared they would upset the chimpanzees. Senate Labor and Human Resources Committee, Americans with Disabilities Act (to accompany S. 933), S. Rep. 116, *supra* n. 60, August 30, 1989, p. 7.

66. *Id.*, p. 22.

67. "A person with lung disease will have a substantial limitation in the

major life activity of breathing." *Id.* See Gerben v. Holsclaw, 692 F. Supp. 557 (E.D. Pa. 1988) (cystic fibrosis is "clearly" a handicapped status under the Rehabilitation Act of 1973); Dept. of Ed., Hawaii v. Katherine D., 531 F. Supp. 517 (D. Hawaii 1982) (cystic fibrosis is a handicap under the Education for All Handicapped Children Act of 1975).

68. S. Rep. No. 101-116, *supra* n. 60, at 7, 22, 24. The *Arline* court, 107 S.Ct. 1123 (1987) cited remarks of Senator Mondale discussing a case in which a woman "crippled by arthritis" was denied a job *not* because she could not work, but because college trustees thought "normal students shouldn't see her." 118 Cong. Rec. 36761 (1972). See Doe v. Region 13 Mental Health– Mental Retardation Commission, 704 F.2d 1402 (5th Cir. 1983) (mental illness qualifies as a handicap under the Rehabilitation Act of 1973).

69. The physical or mental impairment must substantially limit a major life activity. Persons with minor or trivial impairments, such as a simple infected finger, are not disabled within the meaning of the act. S. Rep. No. 101-116, *supra* n. 60, p. 22–23. However, if a defendant discriminates because she regards or perceives the genetic condition as more serious than it actually is, the person is protected under the third prong of the definition.

70. The Senate Labor and Human Resources Committee cited the example of a severe burn victim as a disabled person under the ADA. *Id.*, at 24. Technically, the effects of disfigurement on others could be classified under the third prong of the definition.

71. School Board of Nassau County, Fla. v. Arline, 480 U.S. 273, 318 (1987). While the *Arline* court was concentrating on infectious disease, its conclusion is equally applicable to genetic discrimination: "It would be unfair to allow an employer to seize upon the distinction between the effects of the disease on others and the effects of the disease on a patient, and use that distinction to justify discriminatory treatment," *Id.*, at 282.

72. See Human Genetics Committee of the Council for Responsible Genetics, "Position Paper on Genetic Discrimination," *Genewatch*, May 1990: 3. See further nn. 22–26, 40–44 *supra* and accompanying text.

73. Rep. Augustus Hawkins, *supra* n. 58.

74. Rep. Don Edwards and Rep. Henry Waxman, *supra* n. 59.

75. See Kimmel v. Crowley Maritime Corp., 23 Wash. App. 78, 596 P.2d 1069 (1979) (knee injuries suggesting future harm); Dairy Equipment Co. v. Dept. of Indus., 95 Wis. 2d 319, 290 N.W.2d 330 (1980) (fear of exacerbated future injury from a fall due to having only one kidney); Neeld v. American Hockey League, 439 F. Supp. 459 W.D.N.Y. 1977) (future harm to player with one eye). But see Burgess v. Joseph Schlitz Brewery Co., 259 S.E.2d 248 (S. Ct. N.C. 1979) (disability refers to present, noncorrectable loss of vision, not potentially disabling conditions, so that correctable glaucoma was not a handicap under the state statute).

76. Dairy Equipment Co. v. Dept. of Industry, 290 N.W.2d 330, 335 (S. Ct.

Wisc. 1980) (employee who had only one kidney was "handicapped" within the meaning of the state Fair Employment Act).

77. State Div. of Human Rights v. Xerox Corp., 48 N.E.2d 695, 698 (Ct. App. N.Y. 1985).

78. See Kimmel v. Crowley Maritime Corp., 596 P.2d 1069 (Ct. App. Wash. 1979) (knee injury that might pose future safety risk at sea).

79. Opinion of Charles J. Cooper, assistant attorney general, Office of Legal Counsel, for Ronald E. Robertson, general counsel, Department of Health and Human Services, June 23, 1986.

80. The Justice Department itself reversed its opinion in 1988. Memorandum for Arthur B. Calvahouse, Jr., counsel, re: Application of Section 504 of the Rehabilitation Act to HIV-Infected Individuals, Sept. 27, 1988.

81. See, e.g., Civil Rights Restoration Act of 1987, Pub. L. No. 100-259, Para. 557 (March 22, 1988); ADA, House Report (Education and Labor Committee) No. 101-485 (II) at 58.

82. See, e.g., Doe v. Centinela Hospital, 57 U.S.L.W. 2034 (C.D. Cal. 1988).

83. Fact sheet accompanying letter from Alexander Morgan Capron, Robert Mullan Cook-Deegan, and Edmund Pelligrino to Senator Tom Harkin, June 25, 1990.

84. Examples of discrimination against heterozygotes are provided in notes 38–39 *supra* and accompanying text.

85. See School Board of Nassau County, Fla., v. Arline, 480 U.S. 273, 284 (1987) ("society's accumulated myths and fears about disability are just as handicapping as are the physical limitations that flow from impairment").

86. See L. Gostin, "The Future of Public Health Law," *Am. J. Law & Med.* 12 (1987): 461; Burris, "Rationality Review and the Politics of Public Health," *Villanova L. Rev.* 34 (1985): 933.

87. Vickers v. Veterans Administration, 549 F. Supp. 85 (W.D. Wash. 1982).

88. See "Football Player Sues to Be Permitted to Play," *USA Today*, October 5, 1990.

89. See Neeld v. American Hockey League, 439 F. Supp. 459 (W.D.N.Y. 1977) (denial of plaintiff with sight in only one eye an opportunity to play professional hockey would result in irreparable harm).

90. Peoples v. City of Salina, Kansas, 1990 U.S. Dist. LEXIS 4070 (March 20, 1990). Employers, however, can take action if the person's condition renders him unqualified for the job or would pose a direct threat to others in the workplace. ADA sec. 101(3), 103(b).

91. Letter from Nachama L. Wilker and Ruth Hubbard of the Council for Responsible Genetics to the Honorable Steny Hoyer, U.S. House of Representatives, June 27, 1990.

92. Pub. L. No. 101-336, sec. 102, 202.

93. Title I requires qualification standards, employment tests, or other selec-

tion criteria to be "job-related" and "consistent with business necessity." *Id.*, sec. 102(b)(6). Title II requires the disabled person to meet the "essential eligibility requirements" for the receipt of services or participation in programs or activities.

94. ADA, sec. 103(b). "Direct threat" means a significant risk to the health or safety of others that cannot be eliminated by reasonable accommodation (sec. 101(3)).

95. *Id.*, sec. 102(b)(5).

96. *Id.*, sec. 201(2).

97. *Id.*, sec. 102(b)(5)(A). See Southeastern Community College v. Davis, 442 U.S. 397 (1979) ("undue financial and administrative burdens" or requires "fundamental alteration in the nature of the program"). "Undue hardship" is carefully defined in sec. 101(10) as "requiring significant expense" when considered in light of many enumerated factors.

98. Rep. Henry Waxman, Conference Report on S. 933, July 12, 1990, 136 Cong. Rec. H4626. See Rep. Augustus Hawkins, H4614; Rep. Don Edwards, H4626.

99. See, e.g., Joint Explanatory Statement of the Committee of Conference, paras. no. 2 ("direct threat") and 13 ("health and safety"); Report of the Committee on Labor and Human Resources, *supra* n. 60, at 27.

100. House Conference Report No. 101-596, July 12, 1990 (to accompany S. 933), p. 11. In the House the standard of "direct threat" was extended by the Judiciary Committee to all individuals with disabilities and not simply those with contagious diseases or infections. House Report (Judiciary Committee) No. 101-485 (III), May 15, 1990 (to accompany H.R. 2273), p. 51.

101. Joint Explanatory Statement of the Committee of Conference, para. 10 ("pre-employment inquiries").

102. See School Board of Nassau County, Fla. v. Arline, 480 U.S. 273, 285 (1987).

103. ADA, sec. 103(b) ("Qualification Standards").

104. See Joint Explanatory Statement of the Committee of Conference, para. 10.

105. ADA, sec. 101(3), 103(b).

106. See, e.g., House Report (Judiciary Committee) No. 101-485 (II), May 15, 1990 (to accompany H.R. 2273), p. 51.

107. The legislative record is replete with statements rejecting decision making based upon ignorance, misconceptions, and patronizing attitudes. See, e.g., Senate Report (Labor and Human Resources Committee) No. 101-116, Aug. 30, 1989 (to accompany S. 933), p. 27; House Report (Judiciary Committee) No. 101-485 (III), May 15, 1990 (to accompany H.R. 2273), pp. 52, 153; House Report (Energy and Commerce Committee) No. 101-485 (IV), May 15, 1990 (to accompany H.R. 2273), p. 38; House Report (Educa-

tion and Labor Committee) No. 101-485 (II), May 15, 1990 (to accompany H.R. 2273), pp. 7, 121.

108. Hall v. U.S. Postal Service, 857 F.2d 1073, 1079 (6th Cir. 1988), quoting *Arline*. See also Mantolete v. Bolger, 757 F.2d 1416 (9th Cir. 1985); Strathe v. Dept. of Transportation, 716 F.2d 227 (3rd Cir. 1983).

109. See Peoples v. City of Salina, Kansas, 1990 U.S. Dist. LEXIS 4070 (March 20, 1990).

110. See nn. 88–89 *supra* and accompanying text.

111. B. Tucker, "The EEOC's Safety Defense under Title I of the ADA: Valid or Invalid?" *Nat'l Disabil. Law Rptr.* 2, no. 5 (October 10, 1991): 1.

112. See Jackson v. Johns-Manville Sales Corp., 781 F.2d 394 (5th Cir. 1986) (whether an individual exposed to asbestos but not currently symptomatic can recover today for the likelihood of future cancer).

113. ADA, sec. 101(3), 101(9), 102(b)(5).

114. ADA, sec. 101(9)(A).

115. Vickers v. Veterans Administration, 549 F. Supp. 85 (W.D. Wash. 1982).

116. ADA, sec. 101(10), 102(b)(5A).

117. ADA, sec. 101(10)(B). The Congress specifically rejected the contention of the Supreme Court in TWA v. Hardison, 432 U.S. 63 (1977) that an employer need not expend more than a *de minimus* amount for the accommodations. See Senate Labor and Human Resources Committee (to accompany S. 933), Aug. 30, 1989, at 36.

118. *Id.*, at 35.

119. Pub. L. No. 101-336, sec. 102(b)(7),(8).

120. See Senate Committee on Labor and Human Resources (to accompany S. 933), August 2, 1989, p. 39.

121. For the purpose of the ADA, drug testing is not considered a medical examination, and employers are not prohibited from taking action against a person who is currently using drugs. Pub. L. No. 101-336, sec. 104, 510; but para. 504 regulations (45CFR para. 84.14) limit preemployment inquiries, and at least one federal court (Doe v. Syracuse School District, 508 F. Supp. 333) upheld them in holding an employer in violation of para. 504 for inquiring about a job applicant's prior mental illness.

122. Pub. L. No. 101-366, sec. 102(c)(2)(B).

123. *Id.*, sec. 102(c)(3).

124. *Id.*, sec. 102(c)(4).

125. Senate Committee on Labor and Human Resources (to accompany S. 933), August 2, 1989, 39–40.

126. The Employee Retirement Income Security Act of 1974 (ERISA) may provide an additional source of law to remedy discrimination based purely on cost factors. Section 510 of ERISA makes it unlawful for an employer to "dis-

criminate against a participant or beneficiary for exercising any right to which he is entitled under the provisions of any employer benefit plan . . . or any right to which such participant may become entitled." 29 U.S.C. sec. 1140 (1982). While ERISA does not require employers to provide benefit plans at all, it does prohibit employers from discriminating against employees because they may disproportionately burden a benefits plan now or in the future. The terms "participant" and "beneficiary" suggest that persons are eligible for relief under ERISA only after they have been hired. The purpose of ERISA is to protect "employees" and the "employment relationship." West v. Butler, 621 F.2d 240, 245–246 (6th Cir. 1980). Indeed, courts have thus far restricted entitlement under ERISA to current or discharged employees. See Liebman, *supra* n. 6, at 87–88. Thus, ERISA may provide a remedy for persons with genetic conditions or predispositions once they are hired.

127. *OTA 1990 Report*, at 181–182.

128. See generally, Liebman, *supra* n. 6, at 84–85; Rothstein, "Employee Selection Board on Susceptibility to Occupational Illness," *Mich. L. Rev.* 81 (1983): 1379.

129. Liebman, *supra* n. 6, at 82.

130. See Hearings before the Committee on Labor and Human Resources and the Sub-Committee on the Handicapped, U.S. Senate, 101st Congress, 1st sess., on S. 933 (May 9, 10, 16, and June 22, 1989).

131. Rep. Augustus Hawkins, S. 933, Conference Report, ADA, Cong. Rec. H4614. "An employer could not discriminate against a carrier of a disease-associated gene because such individual may be at higher risk of having a child with a genetic disease whose care would increase costs for the patient's employer." Rep. Henry Waxman, Americans with Disabilities Act of 1990, S. 933, Cong. Rec. H4626.

132. State Division of Human Rights v. Xerox Corp., 480 N.E.2d 695, 697–698 (Ct. App. N.Y. 1985); Garner v. Rainbow Lodge, U.S.D.C. S.D. Tex., Houston Div. H-88-1705 (1989). Shawn v. Legs Co. Partnership, Sup. Ct. N.Y. Cty., AIDS Lit. Rptr., March 10, 1989.

133. Mosby v. Joe's Westlake Rest. No. 86-5045 (Cal. Super. Ct., San Fran. Cty.).

134. Cronan v. New Engl. Telephone, 41 FEP 1273 (Mass. 1986).

135. Shannon v. Charter Red Hospital, Admin. Complaint, Dallas, Tex., April 28, 1986.

136. Shawn v. Legs Co. Partnership, Sup. Ct. N.Y. Cty., AIDS Lit. Rptr., March 10, 1989.

137. Congressman Augustus Hawkins in the Conference Report on S. 933, Cong. Rec. H4614 (July 12, 1990) said: "Allowing the fact of increased costs to justify employment discrimination would effectively gut the protections of the ADA for individuals with disabilities."

138. ADA, House Report (Education and Labor Committee) No. 101-485 (II), May 15, 1990 (to accompany H.R. 2273), at 261.

139. ADA, sec. 501(c)(1).

140. The Employment Retirement Income Security Act (ERISA) regulates employee benefits (including self-insured plans), effectively preempting the states and specifically leaving insurance regulation to the states. See Metropolitan Life Insurance v. Massachusetts, 471 U.S. 724 (1985). ERISA's exemption of self-insured plans from state insurance regulation is not affected by the ADA. ADA, sec. 501(c)(3). See, e.g., ADA, House Report (Judiciary Committee) No. 101-485 (III), May 15, 1990 (to accompany H.R. 2273), at 107–108 ("Concerns had been raised that sections 501(c)(1) and (2) could be interpreted as affecting the preemption provision of ERISA. The Committee does not intend such an implication"). The problem with ERISA's preemption provision is that self-insured plans cannot be required by states to provide certain benefits or to contribute to risk pools. Since an estimated 60 percent of all covered workers are under self-insured plans, a significant limitation is placed on the states that seek to rectify inequitable coverage.

141. See, e.g., Senate Report (Labor and Human Resources Committee) No. 101-116, August 2, 1989 (to accompany S. 933), p. 29. Nor can employee benefit plans be found to violate the ADA under impact analysis simply because they do not address the special needs of every person with a disability, e.g., additional sick leave or medical coverage. House Report (Education and Labor Committee) No. 101-485 (II), May 15, 1990 (to accompany H.R. 2273), at 261. See Alexander v. Choate, 469 U.S. 287 (1985).

142. Breaking down the barriers to employment opportunity for persons with disabilities would actually contribute to the economy. See ADA, sec. 2(a)(9) (unfair discrimination denies people with disabilities the opportunity to compete on an equal basis and costs the U.S. billions of dollars in unnecessary expenses resulting from dependency and nonproductivity). See generally, Office of Technology Assessment, U.S. Congress, *Medical Testing and Health Insurance* (1988).

143. ADA, sec. 501(c).

144. Congress rejected the Supreme Court's restrictive reading of the term "subterfuge" in Public Employment Retirement Systems of Ohio v. Betts, 109 S. Ct. 256 (1989). The court in *Betts* held that subterfuge required some malicious or purposeful intent to evade. See, e.g., Hawkins, *supra* n. 136; Waxman, *supra* n. 130.

145. See House Report (Educational Labor Committee) No. 101-485 (II), May 15, 1990 (to accompany H.R. 2273), at 82–83, 259–260; House Report (Judiciary Committee) No. 101-485 (III), May 15, 1990 (to accompany H.R. 2273), at 35–36, 108.

146. 42 U.S.C. para. 2000e, *et seq.*

147. See Smith v. Olin Chemical Corporation, 555 F.2d 1283 (5th Cir. 1977).

148. Arizona Governing Comm. v. Norris, 463 U.S. 1073, 1108 (1983) (O'Connor, J., concurring).

149. Wright v. Olin Corp., 697 F.2d 1172 (4th Cir. 1982); Hayes v. Shelby Memorial Hospital, 726 F.2d 1543 (11th Cir. 1984). The Supreme Court in Wards Packing Co. v. Antonio, 109 S.Ct. 2115 (1989) said that the plaintiff has the ultimate burden of proof, and the employer need not demonstrate that the challenged practice is "essential" or "indispensable" to show business necessity. Congress has been trying, so far unsuccessfully, to repudiate this holding.

150. Sickle cell lawsuits have been brought under many other theories which are not germane to employment discrimination. See Williams v. Treen, 671 F.2d 892 (5th Cir. 1982) (if state prison officials had denied treatment to persons with sickle cell anemia it would raise a constitutional issue); Ross v. Bounds, 373 F. Supp. 450 (E.D.N.C. 1974) (complaint by black inmates seeking injunctive relief requiring that all black inmates be routinely examined to determine if they had sickle cell anemia or trait did not state a cognizable claim under the Civil Rights Act); Taylor v. Flint Osteopathic Hospital, 561 F. Supp. 1152 (E.D. Mich. 1983) (finding against plaintiff who urged that black patients were discriminated against by being denied reimbursement for "unnecessary" medical treatments especially helpful for typically black problems).

151. See EEOC v. Greyhound Lines, 635 F.2d 188 (3d Cir. 1980) (an African American employee sued under Title VII alleging that a no-beard policy adversely impacted black workers. The employee had a skin condition particular to African Americans).

152. See Narragansett Electric Co. v. Rhode Island Commission for Human Rights, 118 R.I. 457, 374 A.2d 1022, 1026 (1977) (a sickle cell screen would clearly be discriminatory since a nonracial explanation was not possible). Cf. General Electric Co. v. Gilbert, 429 U.S. 125 (1977) (Justices Brennan and Marshall in dissent argued that under the majority opinion the employer could exclude sickle cell–related disabilities from its disability benefits plan and not violate Title VII).

153. Smith v. Olin Chemical Corporation, 555 F.2d 1283 (5th Cir. 1977). See Peoples v. City of Salina, Kansas, 1990 U.S. Dist. LEXIS 4070 (March 20, 1990) (rejecting claim of racial discrimination when a firefighter with sickle cell anemia was dismissed due to risk of sickle cell crisis because he was not qualified for the position).

154. Muller v. Oregon, 208 U.S. 412, 422 (1908).

155. See the Fair Labor Standards Act, 29 U.S.C. para. 201, *et seq.*, and the Occupational Safety and Health Act, 29 U.S.C. para. 651, *et seq.*

156. Pub. L. No. 95-555, 42 U.S.C. para. 2000e(k) (1988).

157. Women affected in pregnancy, childbirth, or related medical conditions shall be treated the same as others who are "similar in their ability . . . to work."

158. International Union, UAW v. Johnson Controls, Inc., 1991 US LEXIS 1715.

159. The pre-set levels were: (1) where any employee recorded a blood lead level exceeding 30ug/dl during the preceding year; or (2) the work site yielded an air sample during the preceding year in excess of 30ug/dl. International Union, UAW v. Johnson Controls, Inc., 886 F.2d 871, 876 & nn. 7, 9 (7th Cir. 1989), *rev'd*, 1991 US LEXIS 1715.

160. *Id.*, at 901.

161. *Id.*, at 885–901. See Hayes v. Shelby Memorial Hosp., 726 F.2d 1543 (11th Cir. 1984); Wright v. Olin Corp. 697 F.2d 1172, 1186 (4th Cir. 1982) ("the problem presented by a fetal protection policy involved motivations and consequences most closely resembling a disparate impact case").

162. *Johnson Controls*, 886 F.2d at 888, 889–901. See also *Hayes*, 726 F.2d at 1543.

163. *Johnson Controls*, 1991 US LEXIS 1715, *19.

164. *Id.*, at *22.

165. *Id.*, at *32.

166. *Id.*, at *33.

167. See Bertin, "Fix the Job, Not the Worker," *L.A. Times*, Nov. 27, 1989, B-7.

168. Occupational Safety and Health Act (OSHA), 29 U.S.C. para. 655(b)(5) (1988). "OSHA has statutory authority to protect the fetuses of lead exposed working mothers. . . . Harm to fetuses is a material impairment of the reproductive system of parents." United Steelworkers of Am. v. Marshall, 647 F.2d 1189, 1256 n.96 (D.C. Cir. 1980), *cert. denied*, 453 U.S. 913 (1981).

169. Cal. Health and Safety Code, paras. 150, 151, 155, 309, 341 (1990); 1990 Cal. Senate Bill 1008 (Ch. 26); Fla. Stat., para. 385.206 (1989); Iowa Code, para. 136A.2 (1989); Illinois 1990 Public Act 86-1028; La. Rev. Stat. Ann., para. 46.2254 (West 1982); Md. Health-Gen. Code Ann., para. 13-101; 191 Revised Stats. of Missouri (1989); N.J. Stat., para. 26: 5B-3 (1987); 1990 N.Y. Laws 900; Va. Code Ann., para. 32.1-68 (1990).

170. Cal. Health and Safety Code, para. 150 (1990).

171. See, e.g., *Id*; N.J. Stat. para. 26:5B (1987); N.J. Annot. sec. 10:5–12a (West Supp. 1988) (any "atypical hereditary or blood trait").

172. Cal. Health and Safety Code, para. 150 (1990).

173. Fla. Stat. Annot. sec. 448.075 (West 1981) prohibits testing for sickle cell but no other genetic trait or disease. Fla. Stat., para. 385.206 (1989) singles out sickle cell and hemophilia for medical care grants. La. Rev. Stat. Ann. para. 46.2254 (West 1982) has a prohibition on employment discrimination that only includes sickle cell.

174. See 191 Revised Stats. of Missouri (1989) (limited to cystic fibrosis, hemophilia, and sickle cell); 1990 N.Y. Laws 900 (sickle cell, Tay-Sachs, or Cooley's anemia).

175. Cal. Health and Safety Code, para. 150 (1990).

176. N.J. Annot. sec. 10:5–12a (West Supp. 1988).

177. See, e.g., Fla. Stat., para. 448.075 (1981) ("No person, firm, corporation, state agency . . . or any public or private entity shall deny or refuse employment to any person or discharge any person from employment solely because such person has the sickle cell trait").

178. Illinois 1990 Public Act 86-1028, S.B. No. 1466.

179. *Id.*

180. E.g., Fla. Stat. Annot., sec. 448.075 (West 1981) (prohibiting testing for sickle cell).

181. E.g., Fla. Stat., para. 385.206 (1989); Iowa Code, para. 136A.2 (1989).

182. Ill. 1990 Public Act, para. 86-1028.

183. 191 Revised Stats. of Missouri (1989).

184. See generally, D. Nelkin and L. Tancredi, *Dangerous Diagnostics: The Social Power of Biological Information* (New York: Basic, 1989).

185. See L. Gostin, "Public Health Strategies for Confronting AIDS: Legislative and Regulatory Policy in the United States," *JAMA* 261 (1989): 1621.

Human Capital and the Discourse of Control: Comment on Paul

Mitchell G. Ash

As a historian of science and of modern Germany, I strongly believe that discussion of the issues being considered here can benefit from some historical perspective. I think we should ask, first, not what *is* but what *was* eugenics? Was it applied science or naturalized prejudice? As research progresses on the history of eugenics movements in different countries, it turns out not to be as easy to distinguish the two as many once thought it was. Francis Galton, the acknowledged founder of the movement, sought to improve the human (primarily the white) race by what he called "judicious marriages," thus increasing the proportion of "desirable" characteristics, but on a voluntary basis. This later came to be called "positive" eugenics.

Toward the turn of the last century, as Friedrich Weismann's distinction between germ plasm and somatoplasm came to be accepted and Lamarckism was rejected, the idea that heredity could be improved by improving social conditions appeared to be in question. As a result, the focus of eugenical thinking began to shift toward the elimination of traits deemed "undesirable" by preventing their reproduction—hence the designation, "negative eugenics."

It was precisely at the points where "desirable" and "undesirable" traits were listed—or taken for granted—that naturalized prejudice, or ordinary social values, did not somehow infect, but *merged with* applied science. The use of such labeling should make it clear that scientists were no different from other members of their class at that time. Eugenics in that context meant the construction of an applied science program that necessarily imposed the values of one social group on other groups. But so far we are only talking about a program. We need to distinguish here between eugenical *thinking* or *discourse* and eugenical *practice*. Even if certain practices deemed eugenical, such as "euthanasia" and compulsory sterilization, have been taboo since the Nazi era, eugenical *discourse*, especially that of "positive" eugenics, may have persisted under other names. More on this shortly.

What *was* eugenical *practice*—racist pseudo-science, as is usually claimed about Nazi eugenics, or rational population policy? Recent research on the so-called race hygiene movement in Germany, to which Diane Paul has made important contributions, suggests that it was *both*. It has been found, for example, that adherents of "Nordic supremacy" *and* of race improvement for "national efficiency" were members of the German Society for Race Hygiene, which was founded at the turn of the century, and that the latter group, not the former, remained in the majority until 1933. The influence of these proponents of "rational" population policy with Prussian and Reich health officials led in the 1920s to the creation of state-funded genetic and marriage counseling programs dispensing birth control and in some instances even abortion advice, along with educational programs that were deemed progressive at the time and could stand comparison with current programs. All of this was strongly supported and in some instances initiated by members of the Society for Race Hygiene.

During the Depression, as pressure on social welfare programs of all kinds increased because of plummeting state revenues, a voluntary sterilization law was drafted and proposed to the (Social Democratic) Prussian state government in 1932—that is, before the Nazis came to power—as a cost-cutting measure. The Nazis' compulsory sterilization law, one of the first laws promulgated after their acquisition of power, thus did not come out of the blue. And cost control—the need to eliminate so-called useless eaters—not racism alone, was among the chief weapons in the rhetorical armamentarium employed in support of the compulsory sterilization program by physicians and Nazi party propagandists alike. Eugenical physicians and anthropologists participated in drafting the law itself, and in the so-called Hereditary Health Courts created to oversee its implementation and to decide in doubtful or disputed cases. American eugenicists rejoiced at the passage of the compulsory sterilization law—at last, they exulted, someone with political power was taking eugenics seriously.

We need to distinguish between this program of "negative" eugenics and the *Lebensborn* program promulgated by SS chief Heinrich Himmler in 1938, which was a "positive" eugenical program based on frankly racist and at best pseudo-Darwinian assumptions. The extension of the claim that "National Socialism is applied biology" to the murder of the Jews is another matter entirely. But even if we limit ourselves to compulsory sterilization and euthanasia, we must still ask: Where were the scientists, and where was science, in all this?

James Watson has articulated a widespread view that incomplete knowledge was abused in a truly cavalier way by the Nazis. But this implies, first, that more complete knowledge would somehow be less likely to be abused—yet the naiveté of that claim is clear as soon as it is stated. The conventional view implies, second, that what knowledge there was was abused only after it had been gathered—that is, that "good" science could not have been done under an evil regime. Diane Paul's work on Soviet geneticist Nikolai Timofeeff-Rosoffsky, just published in *Scientific American*, and my own work on psychological twin research under Nazism, along with many other studies, are showing that high-quality science could be and was well supported by the regime, so long as it appeared relevant to the eugenical project. In other words, the assumption that "good" science was not done under an evil regime cannot be sustained.

But what sort of science are we talking about? The distinction between science, or basic research, and technology trips lightly from our expert tongues, but it is not at all clear that this distinction is obvious in the public mind. In part this is because technological spin-offs of basic research are what most people actually see of science; but the main source of the seeming confusion is that one of the central aspects of the history of science in this century *is* the increasing merging of science and technology, or, more accurately, the technologization of science, resulting in the increasing dependence of basic researchers on government and industry for resources. Inevitably, scientific discourse itself has been affected by this development.

Specifically, in the case of the Human Genome Project (HGP), we have encountered throughout this symposium a combination already familiar to historians of eugenics—of typological thinking in terms of essences, such as "the human genome" or even "the" gene, and technological thinking of the sort evident in repeated references to genes "making" proteins instead of specifying their structure. It is this melding that was so important to eugenical thinking before 1933, combined as it was, and in my view still is, with the biotechnocratic dream of controlling life.

In the case of psychological twin research, as in Timofeeff-Rossoffsky's work, such potential technocratic applications were rarely made explicit. But on closer examination, it becomes clear that what was being developed were *implicit instruments*—tests for the heritability of particular characteristics or combinations of characteristics that could be refunctioned for real-world use with minimal effort. This is what Bruno Latour has called "technoscience"—basic research developed in such a close, symbiotic relationship with its potential applications, informed so fully by the technological ideal of control, that it is hardly distinguishable in its substance from technology.

My problem is not with the use or abuse of such thinking, but with the discourse itself. That discourse, in my view, is not something neutral that can be well or ill used, but is itself abusive. My fear is not that abuse of knowledge generated by the HGP will somehow change human nature, but rather that "human nature," that is to say, a set of social values accepted uncritically among the class of people doing the science, is already in place and will make such abuses seem like natural acts. At stake here is less the widely feared entry of eugenics by some "back door," but the pervasive combination of the discourse of technological control with that of "quality." This is a mode of thinking, speaking, and doing shared by scientists, practitioners, and consumers alike.

This leads to my final, and perhaps most disturbing, question—is technological thinking of this kind any less abusive in human terms when the "choice" is expressed by consumers rather than suppliers of genetic technologies, or by state health care or insurance-company officials? The concept of free choice has become part of America's civic religion. Those who proclaim a faith in "patient autonomy" and believe that the good intentions of physicians in this regard are enough to ensure "good" choices simply miss the point. For after all, where did the values and beliefs that govern these choices come from? Some social theorists are beginning to claim that the whole idea of "individual choice" is a social construction; it is in any case a historical product. No matter how we stand on that issue, it should be clear that is impossible fully to separate individual and social values, and that some choices, no matter whether they be located in individuals, in families, or in impersonal-seeming institutions, are going to be experienced as coercive whether or not actual coercion is coming from any governmental body.

I come now to a final worry, perhaps best expressed by the image of a "genetic supermarket," in which "customers" could eventually assemble collections of desired characters. David Hull dismissed such fears, and nearly all talk of eugenical selection *by individuals* rather than by governments, as "a yuppie inclination" in his essay (see pages 207–219). From an evolutionary perspective, he is surely right; even if thousands of well-off individuals make such choices, it will not affect the human gene pool as a whole in the foreseeable future. But precisely these consumers are the ones with disproportionate influence on, if they are not themselves members of, the power elite of the industrialized world. And it is to them—to us—that my final question must be addressed: Does it really make a difference if eugenical *practices* are now taboo in government policy, if eugenical *discourse* is still very much alive in all our talk about "quality of life" or even "quality children"—internalized as part of the consumerist value system that unites us more than anything else? If that last question has made you uncomfortable, that was my intention.

Commentary on Paul

Alan I. Marcus

I am pleased to have the opportunity to comment on what I consider an extraordinarily insightful essay. Diane Paul's essay deals point-blank with the nuts and bolts of public policy. She cogently dismisses the concerns of those who foresee a restoration or rebirth of the eugenics of the 1920s and 1930s and warns us instead of a back-door eugenics in which parents seek to give their offspring competitive genetic advantages and ultimately produce identical children, or warns that at least the wealthiest—those with the cash to purchase the most individual genetic services in this genetics supermarket—will produce virtually identical progeny.

Of particular relevance, to me at least, is Paul's caveat against studying the eugenics movement to determine the implications of the Human Genome Project (HGP). It suggests a healthy skepticism about the relationship between the past—the unique cultural notions and themes of the early twentieth century, which were embedded with eugenics—and the present—those unique cultural notions and themes embedded within our myriad activities—and treats them as if they were separable eras. Such a notion would send a chill down the spine of many a historian—it demolishes the old saw about those ignorant of the lessons of the past being doomed to repeat them—but Paul skillfully recognizes and acknowledges the likely discontinuity. This leads her to conclude that statists, those most concerned about the coercive power of government, probably miss the mark. They identify the past and present as nearly identical and so gear up to protect against a future fundamentally similar to the past.

Paul's essay is sensitive to the intellectual perils of the present. It reflects the cultural notions and themes of late twentieth-century

America. Present-day Americans act as if they were individuals rather than members of groups, hardly a staging ground for statism. Even if America in the past was marked by class cleavages, racial conflict, and gender separation, those units offered something; they were after all intense groupings of humans, engendering a permanent sense of group identity and belonging. Any sense of permanent or even long-standing cohesion or cohesiveness is gone. Individuals in the present individuated society are said to suffer anomie; each identifies himself or herself as a victim and demands to be empowered to receive the rights to which he or she is entitled. All act as autonomists; they each desperately seek a sense of belonging, participation, or some other form of self-gratification so long as they retain individual initiative, prerogative, and purview. Alliances among the individuated are unidimensional, single-faceted, and generally quite brief.

An individuated society is clearly one in which the HGP can seem extremely menacing. But Paul's proscription about the relevance of the past to the present also has implications for the future. Do not those who plan the future—the futurecasters—fall into the same trap that Paul so adroitly spells out for those who seek to use the past to understand the present and future? Is not the present simply the future's past? And if that is so, does that not undercut the whole idea of planning, the possibility of controlling, dictating, knowing what will happen? Indeed, humans can only have self-knowledge, knowledge of themselves. The knowledge of others in the future is the province of others in the future.

If attempts to plan the future are apparently not possible, what then is one to make of conferences at which plans for the future are considered and evaluated? For late twentieth-century academics, public policy conferences play a crucial role in our self-esteem. They permit us to complain, object, and condemn—to strut our academic stuff—and as we pursue that activity we gain a sense of connectedness and a feeling of authority, if only momentarily. Yet we have committed ourselves to no long-term course or affiliation—there are no new threats or fears generated to the status quo by such meetings—because basically nothing is decided at gatherings of academics.

But we will have experienced, if only for a distressingly brief period, a sense of belonging with those who agree with us (and who had agreed before we presented our arguments) and a sense of participation for having done the "right" thing—we came down on the "right" side of the issue no matter what we said or did. That our recommendations and policy objectives were ultimately of no consequence matters less than that we met and connected, performing activities that reaffirm to us "our cherished social role." We can then all go home relatively content and confident, believing ours was a job well done. We have signified our importance and experienced community through our own academic enterprise. In late twentieth-century America, self-delusion has become synonymous with self-knowledge.

Communicating Complex Genetic Information

Elizabeth J. Thomson

After working in genetics for sixteen years, I have come to believe that while we genetics professionals make our best efforts to communicate with people who seek our services, there are times when we are ineffective at this communication process.

There are a variety of factors that contribute to ineffective communication. One factor is varying education levels between providers and consumers of genetic services. Even those consumers who have an education level similar to their provider's may have education based in another field of study. It is not natural nor is it comfortable for us to communicate in language different from our own. It is particularly noticeable to us when we attempt to communicate at an education level substantially different from our own. Even when we make genuine attempts to communicate at a level different from our own, we still are sometimes far away from the level of understanding of our clients. It is difficult to find simple, straightforward, understandable terms at a sixth-grade reading level to describe deoxyribonucleic acid, genetic mutations, amino acid sequences, protein synthesis, and restriction fragment length polymorphisms. Another factor that significantly interferes with communication is the timing of the discussions. How difficult it is for people to comprehend and remember technical information when the family is in the midst of a crisis, and they are mired in grief, shock, denial, and anger.

Communication is hampered when the consumer has a vision, hearing, learning, or other disability, and there are very few genetic service providers who themselves have any sort of disability that would heighten sensitivities to these individuals.

Communication is further hindered if the people with whom we are trying to communicate do not speak English, or we as providers are not multilingual. It is hampered even more by a virtual absence of ethnocultural variation among genetics service providers in the culturally diverse United States.

To demonstrate this point, I want to relate a story that a New York City anthropologist, Dr. Rayna Rapp, told recently at a meeting that I attended.[1] She told of an African American inner-city woman who had given birth to a child with Down syndrome. This woman was a single mother of three. Her pregnancy had been unplanned. She was, by any standard, poor. Prior to delivery, she had planned to give her baby up for adoption. After the child's preterm birth and the diagnosis of her genetic condition, the mother asked the adoption social worker what kind of family would adopt a child with Down syndrome. The social worker told her that many families from the rural part of the country were interested in adopting children with such disabilities. The woman then asked, "Why?" The well-intentioned, white, urban, middle-class social worker, believing that it would be of comfort to this woman, stated that in such a setting, it was likely that her son would one day be able to learn how to be a farm worker and perhaps have an opportunity to live a productive life. The woman subsequently decided to take her baby home. When questioned regarding her change of mind, she responded, "What that social worker described to me about my boy's future sure sounded like slavery to me."

Some genetics service providers are unaware of or insensitive to the impact that their own value system, which has developed within the context and culture of the biomedical sciences, has on how they communicate genetic information to people. The values that providers place on life with a genetic disorder or some type of disability can influence how they describe a disorder to an individual or a family. In spite of our goal of providing nondirective counseling, it is practically impossible to totally disregard what we, as people, feel about the severity of a disorder. We, too, are influenced by our own life experiences, whether they are personal or professional.

In a recent commentary titled "Twice-told Tales: Stories about Genetic Disorders," Drs. Abby Lippman and Benjamin Wilfond state that they have independently observed that for at least two disorders, Down syndrome and cystic fibrosis, the information provided to individuals considering prenatal testing for these disorders is strikingly different from that provided to parents of a child born with one of these disorders.[2] They state that the information presented about the disorder before testing is largely negative, with much emphasis placed on the technical matters of the testing, while after the birth of a child affected with one of these disorders, the information shared is much more positive. While many will justify the reasons for these differences, the authors believe that it is important simply to acknowledge that these differences exist. They believe that the stories told about genetic disorders are very much affected by who the storyteller is and what his or her own experiences have been. The authors do not claim that one story is better than another, but do question who should be involved in the development of these stories. How many of us have asked people who are themselves directly affected with a particular disorder to help us write the stories we tell?

Another concern I would like to raise is in relation to our ability to obtain truly informed consent *or refusal* for genetic testing. A recent study by a UCLA anthropologist, Dr. Nancy Press, described the information she observed women being given, and also what the women stated they understood, about a prenatal screening test, MSAFP (maternal serum alpha-fetoprotein). Her study was carried out in California, which has mandated that *all* women be offered this screening test during pregnancy. The test itself, however, is voluntary. She stated that although the pamphlet about the test, which had been developed for the California program, was supposed to have been developed for people at the sixth-grade reading level, it was a densely worded nine-page booklet that spent, for example, half a page differentiating between the weeks of pregnancy during which the test *could* be done versus the weeks of pregnancy the test *ought* to be done.[3]

I also want to take a moment to discuss the difference between obtaining consent and allowing a person to choose whether to agree to participate in research or have genetic testing. I would like to share a personal story related to the consent/choice issue. When I was in labor with my daughter, I was asked to participate in three research projects at the hospital in which I delivered. Not surprisingly, I agreed to participate in all three, one of which resulted in my receiving no anesthesia during the last two hours of my labor and delivery. I had been followed in this hospital's obstetrics clinic for eight months prior to my delivery, and these were ongoing research projects—yet never once were these research projects discussed with me. In retrospect, if I had

had an opportunity to read the protocols and consider whether I wanted to participate in these projects, I believe that I might have declined to participate in at least one of them.

Webster defines consent as compliance in or approval of what is done or proposed by another, or compliance with what is requested or desired. The dictionary states that choice suggests the power of choosing, or the opportunity or privilege of choosing freely. As a provider, one must be cognizant of the difference between consent and choice and realize the power that one has over consumers. Even well-educated, outspoken, articulate consumers can turn out to be vulnerable at certain times in their lives, but think of how truly vulnerable consumers are when they are economically disadvantaged or uneducated and have few choices regarding who their providers are and what services they receive. Realize, also, that the majority of consumers in each group like to be viewed as "good patients."

Sometimes providers of genetic services use directive language in their counseling. Please know that I am differentiating between directive language and directive counseling. To refer again to the Press study, the oral discussion regarding the MSAFP test, she observed, often included statements such as, "It is just a simple blood test," or "It won't hurt you or the baby," or "We do recommend the test be done, although you have the option not to have it." She also stated that the language of the results of the MSAFP screening test also seemed counterintuitive, with a *positive* test result being the result a person does not want to get. She pointed out the further potential for confusion in the interpretation of results, because there is a positive high result, a positive low result, and a *negative* result which ends up being the good news. One particularly disturbing finding from this study was the fact that of the women surveyed, more than 30 percent of the women consenting to the test stated that they had had this screening test done because they thought "the state had mandated it."

Another issue I want to raise is in regard to our choice of language in communicating genetic information to families. People with genetic disorders or disabilities are commonly described as "afflicted with, victims of, suffering from, crippled with, or handicapped by . . ." the disorder.[4] We use the names of syndromes as if they were adjectives to describe the person: "a Down syndrome child," "the Marfan syndrome family," or, more recently, "the p53 gene kindred." We sometimes blame the persons with the disorder for *passing on* the disorder to their

children, for *failing* to demonstrate the presence of a genetic mutation, or for *not responding* to the gene therapy.[5] We also sometimes
place a value on a particular risk: "a *low* risk of 2–3 percent," "a *high*
risk of one in four," or "there is *only* a one in a hundred chance that
this could happen again to you."

There has been recent evidence published that women and minorities do not always get treated with the same vigor or in the same manner as white males, at least in the area of treating heart problems with
cardiac bypass surgery. Now I know that this is hard for some to believe, but in the genetics community I have heard of the great strides
in genetics that will influence the future of *man*kind, and I have heard
some of my colleagues talk of gaining a better understanding of chromosomes in *man* or mapping *man*'s genes. We have a reference book
that every geneticist uses on a daily basis. It is called *Mendelian Inheritance in Man*. No book, as far as I am aware, has yet been written
and titled *Mendelian Inheritance in Women*. I sometimes find it surprising that the Human Genome Project was named the *Human* Genome Project. Perhaps there is still hope.

A final issue that I would like to raise is the use of the terminology
"children with special needs." A colleague of mine (who happens to be
blind) recently wrote to me expressing her concerns regarding terminology I had used in a draft of an NIH workshop report on women's
issues and reproductive genetic testing. I had included a statement
about the fact that such testing sometimes allowed a woman to prepare for the birth of a child with special needs. I'd like to read you the
paragraph she wrote regarding the term "special needs." She stated,
"Although this is a commonly used phrase, it perpetuates some of the
problems raised by the whole issue of prenatal diagnosis. In truth,
what is really being talked about is preparing for a child with needs
that, based on a particular characteristic [note her use of the term
"characteristic" rather than "disorder" or "disability"], can be described, named, anticipated in advance." She further stated, "Every
child has special needs. What is being talked about here are the needs
people think they can name and anticipate. They may be right, they
could be wrong, but they are needs that are anticipated because of the
presence of a set of characteristics which suggests those needs to
the minds of parents and professionals. The term 'special needs' as
used here is really an unfortunate euphemism for the term 'disability.'
There is nothing wrong with using the actual word. . . . Disability need
not be shameful."[6]

Now I realize that some of you will tell me that other members of the community with disabilities will disagree with my colleague's criticism, but I want you to know that criticisms of our use of such language should be carefully considered and not disregarded as unimportant or as coming from people who are outside the mainstream. In recent years I have come to realize that one of our problems is that we spend a lot of time talking to each other and to the consumers who like us and think we do a great job at what we do. One of our real challenges for the future is to spend time with those consumers who think we have not done such a great job and carefully consider their comments, criticisms, and suggestions.

We are reaching a point of being able to make predictions about people that have never before been possible. We are now beginning to make these predictions even before the person is born. We are reaching the point of being able to reduce people to their genetic makeup or composition. It will take genuine efforts on the part of professionals and the public to ensure that the whole person continues to come first and not his or her disability, genetic disorder, or genetic composition.

NOTES

1. Rayna Rapp, "Reproductive Genetic Testing: Its Impact on Women," *Fetal Diagnosis and Therapy* (in press).

2. Abby Lippman and Benjamin Wilfond, "'Twice-told Tales': Stories about Genetic Disorders," submitted to the *American Journal of Human Genetics*.

3. Nancy Press, "Collective Fictions: Similarities in the Reasons for Accepting MSAFP Screening among Women of Diverse Ethnic and Social Class Backgrounds," *Fetal Diagnosis and Therapy* (in press).

4. "People First: Sensitive Thinking and Writing for Pediatric Psychology," *Journal of Pediatric Psychology* 16 (1991): 135–136.

5. M. E. Knatterud, "Writing with Patients in Mind: Don't Add Insult to Injury," *AMWA Journal* (February 1991): 10–16.

6. Adrienne Asch, personal letter.

Problems Translating Laboratory Information for High School Students: Are We Doomed?

Kevin Koepnick

The problem is very real. P. D. Hurd summarizes it by saying, "Students are completing high school scientifically and technologically illiterate—foreigners in their own culture."[1]

Foreigners, indeed. Increasingly, those lacking an understanding of the basic processes of heredity and molecular biology are becoming foreigners in their own species, if not their own bodies.

It's all so new. So different. Forty years ago, the structure of DNA was a mystery. Thirty years ago, we began to learn about base codes. In the seventies, it became possible to read the base sequence of DNA. Then came the explosion. In the past ten years, we have learned to remove genes from one species and place them into another. We have learned to mass-produce large quantities of a desirable gene from minute amounts of an original sample. We know the precise reason for many genetic diseases. Business has jumped into biotechnology on a grand scale. Now, we are beginning to read the DNA message stored for millennia in our chromosomes.

This is basic information. Taxpayers need to know what benefits can come from supporting the Human Genome Project to the tune of $200 million per year.[2] Consumers need to know the benefits and risks of new products. Citizens need to understand the implications of such techniques as genetic fingerprinting and prenatal diagnosis.

High schools are the places charged with the preparation of future taxpayers, consumers, and citizens. In fact, the last science class taken by most American students is biology. We ought to have a chance to prepare a majority of young people for a future of genetic and biotechnological advances. But the old method of curricular change by content update will no longer work. Simply teaching the current information and methods of molecular biology seems pointless, since the information and methods are in a constant state of change. We must, instead, improve the scientific literacy of students.

The American Association for the Advancement of Science outlines the dimensions of scientific literacy, in general terms, in the summary to its report *Science for All Americans*. They are as follows:

Be familiar with the natural world and recognize both its diversity and its unity.

Understand key concepts and principles of science.

Be aware of some of the important ways in which science, mathematics, and technology depend upon one another.

Know that science, mathematics, and technology are human enterprises, and know what that implies about their strengths and limitations.

Have a capacity for scientific ways of thinking.

Use scientific knowledge and ways of thinking for individual and social purposes.[3]

Other attempts at sparking curricular change have come up with similar results, most in the form of lists of requirements that must be met for a person to be "literate" in science. Although no single definition of scientific literacy prevails, the many existing definitions share several features. First, all definitions are motivated by the concern that schools prepare all students to function well in an increasingly technology-oriented society. Second, all definitions reflect the notion that students should become acquainted with a fundamental set of science facts and concepts. Finally, they imply that, for most students, the meaning and utility of science will develop not as a result of the memorization of facts, but as a function of understanding the importance and meaningfulness of science in a broad human context.[4]

Certainly, citizens in a technological society ought to be literate in science, but it appears that our attempt to produce this literacy in high school students is doomed to failure. The evidence available shows that most students, even after they have graduated from high school, cannot really understand the science they have been taught, let alone understand how that science relates to technology and society. The reason is simple: what the characteristics and goals mentioned above have in common is that they require students to apply formal reasoning to problems.[5]

Jean Piaget describes intellectual development in terms of four stages: sensory-motor, pre-operational, concrete operational, and formal operational. The last two stages are of interest to high school science teachers since children are expected to begin formal operations at about age eleven, and complete their intellectual development by age fifteen. The formal stage is characterized by the ability to make abstractions.[6] The child at the concrete operational stage, on the other hand, is able to deal only with the concrete reality of a problem and to make only limited extrapolations from it.[7]

No problem, right? Students are able to reason formally at age fifteen, just about the same time they take high school biology. Unfortunately, the ability to think formally does not necessarily translate into the exclusive use of formal thought patterns. In fact, a recent review of research in the development of reasoning among college biology students shows that as many as 50 percent of first-year college students taking biology seldom, if ever, use formal reasoning skills.[8]

So, we seem to be doomed. The idea of promoting scientific literacy within the context of modern genetic technology in a population of concrete thinkers seems an impossible task. Imagine asking a teenager who has difficulty with abstract thought to examine the social and moral implications of inserting portions of human DNA into plasmids, and then transforming bacteria to produce the human gene product. The traditional lecture-discussion format simply does not have a chance.

Yet, perhaps we are not quite doomed. The evidence suggests that the transition from concrete to formal reasoning can at least be eased, if not hastened. One study states that "the concrete student can become formal by careful experimentation and critique."[9] Another study suggests that although formal concepts are not really accessible to students who are not formal thinkers, it is possible to enable students at the concrete level to acquire "surrogate concepts" that can substitute for the real thing, enable them to handle many problems associated with the concept, and make the transition from the surrogate to the real fairly easy at some later time. This study states that it is essential to provide extensive experience with concrete props that model the abstract concept.[10]

The place for exploring, explaining, and criticizing concrete experiences is the laboratory. The problem is that most lab manuals still resemble cookbooks full of tried-and-true recipes.

A different approach is promoted by the National Science Teachers

Association, an approach melding together science, technology, and society (STS). In one STS approach, students may be asked what they know about a biotechnology issue. Then they would be asked what they would like to know about the issue, and what they think they, as adults, ought to know. Simple, inquiry-oriented questions like these begin to make issues personal for students. The job of the teacher, then, is to provide activities which contradict and replace misconceptions, and address students' questions and concerns.

In many cases, the teacher is almost out of the way, since the class feels that it has helped to determine the material to be learned. The key is using real-life problems. Students need to know why they need to know. When they do, they will learn, and learn quickly.

The STS approach to teaching helps with student motivation, but we still run into the problem of abstraction with concrete thinkers. One new piece of curriculum does appear to do an excellent job of fostering formal reasoning. The Biological Sciences Curriculum Studies (BSCS) module *Advances in Genetic Technology*[11] addresses the concepts of biotechnology, important ideas requiring a high degree of abstraction. This module asks students to work with familiar objects acting as models for DNA, RNA, and protein molecules. Work with models will help some students move toward formal operations while providing others with surrogate concepts that will prove useful until more formal reasoning is attained.

Ten years ago, the implications of genetic advances were the stuff of science fiction. Never has the fiction become fact so quickly, nor has it become such an intimate part of our lives. Preparing a scientifically literate citizenry, a citizenry capable of evaluating the implications of molecular biology and biotechnology, is a daunting task with difficult barriers to cross. But we know what does not work, and we are starting to piece together approaches that enhance thinking skills within a contextual framework which includes science, technology, and society.

NOTES

1. J. J. Lagowski, "Literacy," *Journal of Chemical Education* 64 (1987): 733.
2. Robert Shapiro, *The Human Blueprint: The Race to Unlock the Secrets of Our Genetic Script* (New York: St. Martin's Press, 1991), p. 269.

3. F. James Rutherford and Andrew Ahlgren, *Science for All Americans* (New York: Oxford University Press, 1990).

4. A. L. Mitman, J. R. Mergendoller, V. A. Marchman, and M. J. Packer, "Instruction Addressing the Components of Scientific Literacy and Its Relation to Student Outcomes," *American Educational Research Journal* 24 (1987): 611–633; N. C. Harms and R. E. Yager, *What Research Says to the Science Teacher*, vol. 3 (Washington, D.C.: National Science Teachers Association, 1981); National Science Teachers Association, "Science-Technology-Society: Science Education for the 1980s" (Washington, D.C.: NSTA, 1984).

5. Anton E. Lawson, "The Development of Reasoning among College Biology Students: A Review of Research," *Journal of College Science Teaching* 21 (May 1992): 338–344.

6. E. Shahn, "On Science Literacy," *Educational Philosophy and Theory* 20 (1988): 42–52.

7. D. W. Beistel, "A Piagetian Approach to General Chemistry," *Journal of Chemical Education* 52 (1975): 151–152.

8. Lawson, "The Development of Reasoning among College Biology Students."

9. Beistel, "A Piagetian Approach to General Chemistry."

10. J. D. Herron, "Piaget for Chemists," *Journal of Chemical Education* 64 (September 1987): 733.

11. E. Drexler, M. A. W. Hinchee, D. T. Lundberg, L. B. McCollough, J. D. McInerney, J. Murray, and R. Storey, *Advances in Genetic Technology* (Lexington, Mass.: D. C. Heath and Co., 1989).

Reflections on the Law and Ethics
of the Human Genome Project

Peter David Blanck

The symposium session "Ethical and Legal Implications of the Human Genome Project" raises important issues about the potential impact and scope of the HGP. I will comment on the essays by Professors Walters and Gostin presented at the session from my perspective as a lawyer, social scientist, and former chair of the American Psychological Association's Committee on Standards in Research (CSR). My comments on Walters' essay (see pages 220–231) raise issues related to the protection of vulnerable groups in society, such as children, patients, persons with disabilities, and persons in poverty, when they participate in human genome research, therapy, or intervention. My comments on Gostin's essay (pages 122–163) relate to the concept of genetic discrimination, its potential coverage by the Americans with Disabilities Act of 1990 (ADA), and my relevant empirical research.

INFORMED CONSENT, PRIVACY, AND CONFIDENTIALITY

Walters raises issues central to informed consent, privacy, and confidentiality in human genome research, diagnosis, and therapy. Several of these issues are particularly relevant to legal and ethical questions in human subjects research on vulnerable populations addressed by the CSR. The primary purpose of the CSR was to provide advice on issues and standards related to the protection of human participants in psychological and behavioral research.[1] The central question raised here relates to the rights of vulnerable populations who may participate in or be excluded from participating in human genome research, therapy, or intervention.[2]

There are special ethical and legal issues associated with genetic research, therapy, and intervention involving vulnerable populations. Vulnerable persons are often disenfranchised from society and have little voice in research or intervention development or regulation. It is unclear whether the ADA alone will guarantee vulnerable populations with asymptomatic genetic disorders the right to receive equal opportunity for health insurance and medical care. Ironically, with increased exposure to unemployment, health risk, substance abuse, and abuse and neglect, vulnerable populations are the focus of increased research efforts and legal protection. Yet little systematic attention has been devoted to the ethical and legal rights of vulnerable persons as a distinct subgroup of genetic research and intervention participants.

A list of legal and ethical dilemmas related to genetic research, therapy, and intervention that involve vulnerable populations may be formulated through the following questions:

What legitimate role do parents or guardians of persons with mental and genetic disabilities, or of those who are adjudicated to be incompetent, have in the consent process?

What legal and ethical guidelines should be followed by the parents of children with genetic disorders in regard to decisions about long-term familial involvement in genetic research and intervention?

What constitutes valid informed consent (versus coerced participation) for research or intervention involving persons in poverty with genetic disorders when monetary compensation is provided?

What are the independent responsibilities of researchers to the children of parents who consent to participate in genetic studies?

What role will institutional review boards (IRBs) play in the monitoring of longitudinal genetic research protocols?

What legal and ethical responsibility or duty of care do researchers or others have to warn third parties (e.g., spouses, employers, insurers) of potentially harmful genetic facts relevant to the potential offspring of the subject?

How will courts redress present and future harms (e.g., through the use of monetary and injunctive remedies) resulting from genetic research, intervention, or therapy?[3]

Enhanced dialogue on these issues, as begun by the work of Walters and others, is needed before any conclusive statements can be made about the long-term effects of genetic research, therapy, and intervention on all segments of our society.

GENETIC DISCRIMINATION AND THE AMERICANS WITH DISABILITIES ACT

Gostin raises important issues regarding the relationship between genetic discrimination and the social reform legislation embodied in the ADA.[4] Genetic discrimination occurs when, on the basis of genetic information, an individual is denied a right or privilege that is available to others. Despite increased sensitivity to the privacy, consent, or confidentiality rights of citizens, as suggested above, the HGP will result in new forms of information that may be used to predict an individual's present and future employment potential or health risks.

Gostin suggests that empirical study is needed to understand the prevalence of genetic discrimination. In particular, study is needed on the myths and stereotypes in society that may be the source of genetic discrimination by employers, insurers, or the general public. Gostin explores carefully the extent to which the ADA and its judicial and legislative progeny may act as a safeguard against genetic discrimination in the employment relationship and in the provision of insurance and health care benefits. I will discuss briefly these two issues in the context of my ongoing empirical study of the ADA.

My colleagues and I are conducting several longitudinal studies related to the impact of the ADA on the integration of persons with disabilities into the mainstream of society.[5] This empirical research focuses on, among other issues, the employment provisions (Title I) of the ADA and their impact on the benefits, privileges, and opportunities afforded employees with disabilities. The program of research is exploratory, intended to be useful to employers and employees, and meant to provide empirical information on the systematic assessment of compliance with the ADA.

Consistent with Gostin's emphasis, one long-term goal of the project is to replace with hard data the many myths and misconceptions regarding persons with disabilities. Empirical information is gathered longitudinally from the perspective of some three thousand "consumers" of the ADA (primarily persons with mental and physical disabilities) and fifty "users" of the ADA (employers of these persons with disabilities).

The empirical study of consumers of the ADA explores their medical needs, adaptive equipment needs, income levels, insurance needs, skill development, need for accommodations in home settings and the workplace, and their perceptions of barriers to obtaining and retaining equal employment. The study of users of the ADA explores employers' experiences and concerns about hiring and retaining persons with disabilities, providing insurance and benefits to persons with disabilities, and receiving state and local assistance to support the employment of persons with severe disabilities.

Four preliminary findings of the empirical study of the ADA are relevant to the emerging discussion about genetic discrimination in employment and in the provision of universal insurance to employees:

Persons with very complex physical and mental disabilities are being effectively employed, and the cost of accommodating these employees may not be as great as many employers feared.

Persons with disabilities residing in less-integrated living settings tend to earn significantly less income and have less employment opportunity than persons with disabilities residing in more-integrated living arrangements.

Employers are very satisfied with the attendance, productivity, initiative, and dedication to work of their employees with disabilities.

Nearly all employers surveyed do not believe that insurance rates will skyrocket if they hire more persons with disabilities.

These preliminary findings raise more questions than they answer. But they may help develop methods that illuminate the prevalence of genetic discrimination in society and the potential deterrent effect of the ADA. Several empirical problems with regard to the future study of genetic discrimination in employment and in the provision of universal insurance coverage include the following:

How to assess empirically whether a person with a genetic disorder has equal opportunity for employment for which he or she is otherwise qualified?

How to define and assess empirically whether individuals with asymptomatic genetic disorders are substantially limited in major life activities or "regarded" as having a disability for purposes of coverage under the ADA?

How to track longitudinally whether families, minorities, vulnerable populations, or other subgroups in society with prevalent genetic disorders are excluded from employment and equal insurance benefits, opportunities, and privileges?

How to assess the impact that genetic discrimination has on economic opportunity and advancement for vulnerable populations with genetic disabilities?

These and other questions will need to be addressed to ensure that genetic discrimination does not render entire segments of otherwise qualified persons in society unemployable or uninsurable. The 1991 Civil Rights Act begins to help address these issues by establishing a Glass Ceiling Commission designed to study and track the advancement of women and minorities in the workplace.[6] This commission is to address issues of discrimination based on race, ethnicity, and gender, and may ultimately shed further light on the nature and prevalence of discrimination based on genetic disorders. It is likely, however, that clarifying federal legislation will be needed to prevent genetic discrimination under the ADA, absent judicial expansion of the scope of the act.

Persons with disabilities have been excluded from society and have been subjected to deep-rooted prejudices, myths, and stereotypes about their needs and abilities. Persons with genetic disorders have received similar treatment. HGP offers tremendous promise for aiding in the treatment of and research and intervention in genetic disorders. But with the advances of HGP come new ethical and legal questions that require careful attention and that affect all segments of our society. The work of Walters and others is helping to begin the dialogue on these important issues.

The social reform legislation embodied in the ADA similarly holds great promise for promoting equal opportunity for qualified persons with disabilities in all aspects of life. The work of Gostin and others highlights the enormous potential that the ADA may have for helping to eliminate genetic discrimination in employment and in the provision of adequate insurance to otherwise qualified persons. Much more empirical information about the behavior, attitudes, and misconceptions of persons with and without genetic disabilities, their employers, and insurance providers is needed to understand the true societal impact of the HGP.

NOTES

1. Peter David Blanck, Alan S. Bellack, Ralph L. Rosnow, Mary Jane Rotheram-Borus, and Nina R. Schooler, "Scientific Rewards and Conflicts of Ethical Choices in Human Subjects Research," *American Psychologist* 47 (1992): 959–965; Thomas Grisso, Elizabeth Baldwin, Peter David Blanck, Mary Jane Rotheram-Borus, Nina R. Schooler, and Travis Thompson, "Standards in Research: APA's Mechanism for Monitoring the Challenges," *American Psychologist* 46 (1991): 758–766.

2. Peter David Blanck, "On Integrating Persons with Mental Retardation: The ADA and ADR," *New Mexico Law Review* 22 (1991): 259–276.

3. For a comprehensive review, see Lori B. Andrews and Ami S. Jaeger, "Confidentiality of Genetic Information in the Workplace," *American Journal of Law & Medicine* 17(1,2) (1991): 75–108.

4. *Americans with Disabilities Act of 1990* (ADA), Pub. L. No. 101-336, 104 Stat. 327 (codified at 42 U.S.C.A. §§ 12101–12213 and 47 U.S.C. § 225 and § 611 (West Supp. 1990).

5. Peter David Blanck, "Empirical Study of the Employment Provisions of the Americans with Disabilities Act: Methods, Preliminary Findings and Implications," *New Mexico Law Review* 22 (1991): 119–241; Peter David Blanck, "The Emerging Work Force: Empirical Study of the Americans with Disabilities Act," *Journal of Corporate Law* 16 (1991): 693–803; Peter David Blanck, *The Americans with Disabilities Act: Putting the Employment Provisions to Work. A White Paper of the Annenberg Washington Program* (Washington, D.C.: Annenberg Washington Program, Communications Policy Studies, Northwestern University, 1993).

6. *The Civil Rights Act of 1991*, Pub. L. No. 102-166, 105 Stat. 1071 (1991).

Why Fund ELSI Projects?

Robert F. Weir

Approximately 3 to 5 percent of the federal funding for the Human Genome Project (HGP) is going to pay for educational and research efforts focused on the ethical, legal, and social implications (ELSI) of the work being done in molecular genetics. In one respect, this funding is good news: it represents the largest expenditure of money for biomedical ethics and health law in the country, and it enables ethicists, attorneys, and social scientists to have an unprecedented opportunity to do funded research. In other ways, however, the funding brings potential problems: it may divert the research efforts of many professionals from other ethical issues and social problems as significant as those arising in genetics, and it may subject the research of some professionals in ethics, law, and the social sciences to the questions of HGP critics, the skepticism of opponents of applied ethics, and the review of government bureaucrats in ways that will prove uncomfortable.

Simply put, some critics of this expenditure of federal funds will question why such money is being spent on educational and research efforts outside the biomedical sciences. As one person asked in a belligerent manner, "What's the big deal about this ethical and legal stuff?"

I will not try to supply either a complete answer to that question or a persuasive defense of the funding for ELSI projects. I will, however, provide three examples that will suggest why at least some of the ELSI funding is important to us as individuals, members of families, and taxpayers in the country now paying for the HGP.

The first example has to do with biomedical research. All universities in the U.S. performing federally funded research have had institutional review boards (IRBs) since the mid-1970s. Medical researchers, biologists, behavioral scientists, social scientists, and other researchers using human subjects for research purposes have become accustomed to the federal requirement of gaining IRB approval for their research proposals before sending the proposals to a funding agency. However, problems are now arising with regard to molecular genetics research, and most IRBs do not have members on the committees who are sufficiently aware of the work being done in molecular genetics labs, of the risks inherent to research subjects (and, sometimes, their relatives) who provide blood or other tissue for analysis in molecular biology labs, or of some of the difficulties in securing adequately informed consent from potential subjects in genetics research.

The type of knowledge and the nature of concerns about genetics-related issues that should characterize the work of IRBs can be easily illustrated. Conventional consent documents for genetics research typically identify only minimal physical risks to participants (e.g., the risks inherent in obtaining a blood sample). Concerns about genetic privacy (e.g., the inappropriate release of genetic information on individuals) or genetic discrimination (e.g., loss of employment or insurability based on the identification of a genetic characteristic) are rarely, if ever, mentioned. Thus participants may unknowingly agree to participate in a research study, but later run the risk of being denied employment or health insurance if the results of their participation become known to other parties.

Several of us at the University of Iowa have received ELSI funding to address this problem as part of a large genome center grant. We are funded, in part, to examine a large number of consent documents for genetics-related research around the country to ascertain if these documents adequately inform potential research subjects of the projected laboratory use of their blood or other biological materials and if the documents adequately inform the subjects of the possible risks they face when interpretive tests in a molecular lab disclose previously unknown information about them. We then plan to develop new models for consent documents that will accurately reflect the nature of molecular genetics research and the risks inherent to research subjects (and patients in clinical settings). We plan to produce educational materials for IRBs, based on our research findings and our development of one or more model documents, so that they can better protect the interests of both researchers and subjects in research using molecular genetics technologies.

The second example concerns the problems of discrimination addressed in Larry Gostin's essay. As he correctly documents, employers and insurers now have available to them the technologies of molecular genetic analysis that can be used, should they choose to do so, to discriminate against individuals and families based on genetic conditions they have, carry, or may be susceptible to in certain environments. Federal legislation, including the 1990 Americans with Disabilities Act, thus far provides inadequate protection from this new kind of discrimination.

A research project already funded with ELSI money is providing very helpful information regarding state legislative efforts to address this problem. The project, carried out by Jean McEwen and Philip Reilly at the Eunice Kennedy Shriver Center for Mental Retardation in Boston, provides updated analyses of existing and proposed legislation specifically intended to regulate the collection, use, and potential misuse of genetic data about individuals. The March 1992 update of their report provides an analysis and critique of the few existing state laws on genetic discrimination (e.g., in Wisconsin and Iowa), legislative bills proposed in another fifteen states, and the proposed Human Genome Privacy Act currently stalled in Congress.[1] The report also includes their suggestions regarding features that should be included in state laws governing informed consent to genetics research, the confidentiality and privacy of genetic information, and the use of genetic information by employers and insurers.

The third example involves a case of discrimination based on genetics. Last year, in the publication *GeneWATCH*, Jamie Stephenson provided an account of the problems her family has had with the health insurance industry. In the summer of 1991, she and her husband were informed that their family's health insurance had been canceled because two of her sons have been diagnosed with fragile-X syndrome. As the director of the Fragile-X Resource Center of Northern New England, she was appalled that a family could lose its health insurance over a condition causing developmental disability but no medical risk to other persons.

Stephenson was subsequently dismayed to find out that the Americans with Disabilities Act provided no protection for this kind of case, that the insurance company could legally cancel their policy, and that no other private insurance companies would offer them insurance unless the policy excluded the two children with fragile-X syndrome. She concluded with a warning for all of us:

> As the Human Genome Initiative . . . maps more and more of genetic "defects," no one in our society will be safe from discrimination. Families who might participate in HGI studies [may] become reluctant to do so out of fear that information will be "leaked" and discrimination will result. This issue must be addressed aggressively and immediately.[2]

As it happens, a number of other persons share Stephenson's concern about insurance companies and the very real risk that an increasing number of companies will use genetic information to deny health insurance to individuals and families who have genetic conditions that may cause disability and/or require expensive, ongoing medical treatments. To cite two examples, the HGP has a national Task Force on Genetic Information and Insurance, funded with ELSI money; and the University of Florida has a major ELSI grant to analyze a number of complex issues related to the genetic identity of individuals, the use of genetic testing by insurance companies, and the societal risk of having large numbers of persons denied insurance coverage for health problems over which they have no control.

In my view, ELSI funds spent on the development of better consent documents for genetics research, stronger legal protections against genetic discrimination, and more just insurance practices for persons having genetic conditions will be money well spent. I hope that you and most other taxpayers agree.

NOTES

1. Jean E. McEwen and Philip R. Reilly, "State Legislative Efforts to Regulate Use and Potential Misuse of Genetic Information," March 1992, unpublished manuscript.

2. Jamie Stephenson, "A Case of Discrimination," *GeneWATCH* 7 (February 1992): 9.

III. Genders, Races, and Future Generations

Constructs of Genetic Difference: Race and Sex

Ruth Hubbard

Since its beginnings, the science of genetics has been caught up in the dialectic between likeness and difference. When people think about heredity, what they hope scientists will explain is how it is that Johnny has Grandpa's nose and Aunt Mary's chin. But they also want to understand how come little Susie doesn't look like anyone else in the family.

Mendel's experiments, published in 1865, are the conventional way to date the origin of genetics. These experiments grew out of an interest in the breeding patterns of hybrids. The phenomena Mendel was trying to explain had to do with the fact that, for example, red-flowered pea plants, when crossed with each other, usually breed true, but that every so often they produce a plant with white flowers. Mendel's so-called laws of heredity are formal ways to explain how this happens.

Mendel did not concern himself with what went on inside his pea plants. He was describing patterns of transmission of visible "characters" or "traits" between successive generations. Only once in his classic paper did he refer to "factors" inside the plants, which he assumed corresponded to the visible traits.

By around 1900, when Mendel's paper was rediscovered, biologists were less preoccupied with the external appearances of organisms—with taxonomy and anatomy—and had become increasingly interested in how organisms function—in physiology and biochemistry. When the Danish botanist Wilhelm Johannsen coined the term "gene," he was redirecting attention from Mendel's "traits" to Mendel's "factors." But whether they were concerned primarily with traits or with genes, until quite recently geneticists have only been able to explore the results of "mutations"—that is, change and difference. As the British geneticist J. B. S. Haldane wrote in 1942: "Genetics is the branch of biology which is concerned with innate differences between similar organisms. . . . Given a black rabbit and a white rabbit, the geneticist asks how and why they differ, not how and why they resemble each other."[1] The latter question has been the agenda for embryologists.

Even after genes were identified chemically as DNA and the double helical structure of DNA was described during the 1950s, geneticists continued to explore patterns of difference—mutations—not the reproduction of likeness. The only time DNA is involved in the production of likeness is while it itself is being replicated, which happens when cells divide and each of the two new cells ends up with the identical complement of DNA that was present in the original cell.

By studying mutations of biochemical reactions in relatively simple organisms, such as the fungus *Neurospora*, biologists concluded that different genes account for the presence or absence of specific enzymes, which are proteins. And it is important to understand that the transmission of traits from one generation to the next involves not only genes, that is, DNA, but depends on the interplay of genes with proteins. In fact, the entire metabolic apparatus of the cell is involved and, indeed, the whole organism growing and developing within its environment. Genes are embedded in cells, which are parts of organisms, and those cells and organisms constitute the functional realities within which DNA plays its part. Even though genes are often mistakenly referred to as "blueprints" of the organism, the fact is that DNA is in no sense more fundamental or privileged in its contributions to the maintenance or reproduction of a functioning organism than is the rest of the cell or, indeed, the entire organism.

Until the 1970s scientists were able to tell us hardly anything about how similarities are transmitted to successive generations. More recently, since the biochemical language of genes has begun to be translated into the language of protein molecules, molecular biologists have been able to make a few limited statements about the transmission of similarities between successive generations of cells, though not of organisms. The development of an organism always involves a complex interplay between different kinds of cells in which different kinds of genes, proteins, and other metabolites function. It also depends on mutual relationships the organism establishes with its surroundings, which include other organisms.

This is why the claim that scientists can explain how similarities or differences between any but the simplest chemical traits (such as the composition of the proteins responsible for the manifestation of sickle cell anemia or cystic fibrosis) are transmitted to successive generations rests in ideology, not in precise descriptions of biological processes. Basing differences between the so-called races or between women and men, as groups, in "genes" simply uses that status-laden word to legitimate ideological constructs. This is not to say that genes are not involved in the production of skin pigments or genitals. Genes are the functional units of DNA that specify the composition of proteins. And since different kinds of proteins take part in all the biochemical reactions that go on in organisms, genes also are involved. But so are many other metabolites and conditions that must be just right for organisms to develop, be born, grow, and survive.

Yet, it is a fact that scientists have put a good deal of effort into examining the biological basis of various characteristics that have cultural and political significance, including differences between so-called races or between women and men. And they have often made it appear as though differences in power between individuals or groups of people were inevitable and natural results of biological difference, and hence of genes.

This became critically important in the eighteenth century, when support for the aims of the revolutions fought for liberty, equality, fraternity, and for the "rights of man" needed to be reconciled with the obvious inequalities between nations, races, and the sexes. It is well to realize that as late as the sixteenth century, authors described the peoples of Africa as superior in wit and intelligence to the inhabitants of northern climes, arguing that the hot, dry climate "enlivened their temperament,"[2] and two centuries later Rousseau still rhapsodized about the Noble Savage. The industrialization of Europe and North America depended on the exploitation of the native populations of the Americas and Africa. So it became imperative to draw distinctions between that small number of men who were created equal and everyone else. By the nineteenth century, the Noble Savage was a lying, thieving Indian, and Africans and their enslaved descendants were ugly, slow, stupid, and in every way inferior to Caucasians. Distinctions also needed to be drawn between women and men, since, irrespective of class and race, women were not included among "all men" who were created equal.

Although there are many similarities in the ways biologists have rationalized the inequalities between the races and sexes (and continue to do so to this day), our discussion will be clearer if we look at the arguments separately.

Let us begin by looking at the so-called races of man. The writer Allan Chase dates scientific racism from the publication of Malthus' *Essay on Population* in 1798 and argues that it focused on class distinctions among Caucasians rather than on distinctions between Caucasians and the peoples native to Africa, America, and Asia.[3] However, Stephen Jay Gould attributes the first scientific ranking of races to Linnaeus some forty years earlier.[4] Linnaeus went further and arranged the races into different subspecies. He also wrote that Africans, whom he called *Homo sapiens afer*, are "ruled by caprice," whereas Europeans (*Homo sapiens europaeus*) are "ruled by customs," and that African men are indolent and African women are shameless and lactate profusely.

Both Linnaeus and Malthus did their work more than two centuries after the beginning of the European slave trade, which became an important part of the economies of Europe and the Americas. But their work was contemporary with the intellectual and civic ferment that led to the American and French revolutions of 1776 and 1789 and to the revolution that overthrew slavocracy in Haiti in 1791. As the Guyanese political thinker and activist Walter Rodney pointed out, it is wrong to think "that Europeans enslaved Africans for racist reasons." They did so for economic reasons, since without a supply of free African labor, they would not have been able "to open up the New World and to use it as a constant generator of wealth. . . . Then, having become utterly dependent on African labour, Europeans at home and abroad found it necessary to rationalize that exploitation in racist terms."[5]

Physicians and biologists helped to legitimate such rationalizations by constructing criteria, such as skull volume, brain size, and many others, by which they tried to prove scientifically that Africans are inferior to Caucasians. Gould's *Mismeasure of Man* describes some of these measurements and documents their often patently racist intent. Gould also illustrates the ways in which, for example, the distinguished nineteenth-century French anatomist Paul Broca discarded criteria by which white men could not be made to rank highest. And he shows how Broca and the American craniometer Samuel George

Morton fudged and fiddled with their data in order to make the rankings come out as these men knew they must: Euro-American men on top, next Native American men, and then African American men. Women presented a problem: though clearly white women ranked below white men, were they to be above or below men of the other races? A colleague of Broca's addressed this conundrum in 1881. "Men of the black races," he wrote, "have a brain scarcely heavier than that of white women."[6]

In 1854, Dr. Cartwright, an American physician, wrote an article entitled "Diseases and Peculiarities of the Negro," in which he asserted that a defect in the "atmospherization of the blood conjoined with a deficiency of cerebral matter in the cranium . . . led to that debasement of mind which has rendered the people of Africa unable to take care of themselves."[7] And racialist biology did not end with slavery. Writing during World War II, the Swedish economist Gunnar Myrdal marveled that the American Red Cross did not accept African Americans as blood donors. "After protests," he wrote, "it now accepts Negro blood but segregates it to be used exclusively for Negro soldiers. This is true at a time when the United States is at war, and the Red Cross has a semi-official status."[8] The American Red Cross continued to separate the blood of African Americans and European Americans until December 1950, when the binary classification into "Negro" and "white" was deleted from the donor forms. Howard Zinn has pointed out the irony that, in fact, an African American physician, Charles Drew, developed the blood-banking system in the first place.[9]

What can we, at the present time, say about the biology of race differences? Looking at all the evidence, there are none.[10] Demographers, politicians, and social scientists may want to continue using "race" to sort people, but as a biological concept it has no meaning. Human beings (*Homo sapiens*) are genetically a relatively homogeneous species. If Europeans were to disappear overnight, the genetic composition of the species would hardly change. About 75 percent of known genes are the same in all humans. The remaining 25 percent are known to exist in more than one form, but all the forms can be found in all groups, though sometimes in different proportions.[11] Another way to say this is that because of the extent of interbreeding that has happened among human populations over time, our genetic diversity is pretty evenly distributed over the entire species. An occasional, relatively recent mutation may still be somewhat localized within a geographic area, but about 90 percent of the variations known to occur among humans as a whole occur also among the individuals of any one national or racial group.[12]

Another important point is that for any scientific measurement of race differences, we first have to construct what we mean by race. Does the least trace of African origins make someone black, or does the least trace of European origins make someone white? The former definition is more widely accepted, but it is a social convention and not a fact of biology.

The U.S. census for 1870 contained a third category, "Mulatto," for "all persons having any perceptible trace of African blood" and warned that "important scientific results depend on the correct determination of this class."[13] The U.S. census for 1890 collected information separately for "Quadroons and Octoroons"—people one in four of whose grandparents or one in eight of whose great-grandparents were African, "while in 1930, any mixture of white and some other race was to be reported according to the race of the parent who was not white."[14] Finally, in 1970—only about twenty years ago—the Statistical Policy Division of the Office of Management and the Budget warned that racial "classifications should not be interpreted as being scientific or anthropological in nature."[15]

How, then, should we interpret such statistics as that "black men under age 45 are ten times more likely to die from the effects of high blood pressure than white men," that "black women suffer twice as many heart attacks as white women," and that "a variety of common cancers are more frequent among blacks . . . than whites,"[16] especially when some scientists and the media keep stressing the genetic origin of these conditions? Does that not prove that there are inherent, biological differences between blacks and whites, as groups?

The fact is, it doesn't. It is unfortunate and misleading that U.S. health statistics usually are presented in terms of the quasi-biological triad of age, race, and sex, without providing data about employment, income, housing, and the other prerequisites for healthful living. Even though there are genetic components to skin color, as there are to eye or hair color, there is no biological reason to assume that any one of these is more closely related to health status than any other. Skin color ("race") is no more likely to be biologically related to the tendency to develop high blood pressure than is eye color.

To come up with rational explanations, we need to take account of the fact that the median income of African Americans since 1940 has been less than two-thirds that of Americans of European descent. Disproportionate numbers of African Americans live in more polluted and run-down neighborhoods, work in more polluted and stressful workplaces, and have fewer escape routes out of these living and work situations than have Euro-Americans. Furthermore, African Americans at all levels of society experience stress arising from their history and day-to-day experience of discrimination. It is not surprising to find consistent discrepancies in health outcomes between "blacks" and "whites."

The physician Mary Bassett and the epidemiologist Nancy Krieger, looking at mortality risks from breast cancer, have found that the black/white differential of 1.35 drops to 1.10 when they look at African American and European American women of comparable social class, as measured by a range of social indicators.[17] And within each "racial" group, social class is correlated with mortality risk. Because of racial oppression, being black is a predictor of increased health risk, but so is being poor, no matter what one's skin color may be. The fact that even at comparable education and class standing, some health risks appear to be greater for African Americans than for European Americans needs to be analyzed by taking into account the range of factors that constitute the panorama of American racism.[18]

How, then, are we to think about sex? Here I would argue that our society's insistence that any muting of differences between women and men is intolerable exaggerates biological differences and therefore enhances our impressions of them.[19] (Notice our use of the phrase "the opposite sex" instead of "the other sex.") In fact, women and men exhibit enormous overlaps in body shape and form, in strength, and in most other parameters. The diversity within each of the two groups is often as large as, or even larger than, the differences between them. Yet biological differences exist in women's and men's procreative capacities. Our society may exaggerate and overemphasize them, but the fact is that to procreate together, people need to be of two kinds.

The question I want us to look at briefly is to what extent ideological commitment to the differences our society ascribes to women and men in the social and political spheres influences the ways biologists describe the actual biological differences involved in procreation.

To do this, it is useful to start with a quick look at Darwin's theory of sexual selection, which embedded Victorian preconceptions about the differences between women and men in modern biology. Sex is important to Darwin's theory of evolution by natural selection because the direction evolution takes is assumed to depend crucially on who mates with whom. This is why Darwin needed to invent the concept of sexual selection—the ways sex partners choose each other. Given the time in which he was writing, it is not surprising that he came up with the Victorian paradigm of the active, passionate, sexually undiscriminating male who competes with every other male in the pursuit of every available female. By contrast, Darwin wrote, females are passive, coy, and sexually unenthusiastic but choosy, and go for the winner. He theorized that this makes for greater competition among males than among females, and since competition is what drives evolution by natural selection, males are the vanguard of evolution. Females get pulled along by mating with the most successful males.

The essentials of this interpretation have been incorporated into modern sociobiology.[20] Only in the last few years have feminist sociobiologists such as Sarah Blaffer Hrdy revised this canon and pointed out that females, too, are active, sexually aggressive, and competitive, and that males, too, nurture and are passive. Among animals as well as among people, females do not just stand by and wait for the most successful males to come along.[21]

I have criticized the Darwinian paradigm of sex differences elsewhere.[22] The point I want to stress here is that until quite recently, the active male/passive female dyad has been part of biological dogma and is the metaphor that informs standard descriptions of procreative biology at all levels.

For example, the differentiation of the sex organs during embryonic development is said to proceed as follows: the embryo starts out sexually bipotential and ambiguous, but early during embryonic development in the future males, the Y chromosome somehow induces part of the undifferentiated, primitive gonad to turn into fetal testes. The testes then begin to secrete fetal androgens (called "male hormones"), which are instrumental in the differentiation of one set of embryonic ducts into sperm ducts and external male genitalia (the scrotal sac and penis), while another set of ducts atrophies. By this standard account, in the absence of a Y chromosome nothing happens until somewhat later when, without special hormonal input, another part of the undifferentiated fetal gonad differentiates into ovaries and, since the ovaries do *not* secrete androgens, the other set of fetal ducts differentiates into the fallopian tubes, uterus, and vagina, and into the external female genitalia (the labia and clitoris).

Notice that in this description, male differentiation is actively triggered by the Y chromosome and by so-called male hormones; female differentiation happens because these triggering mechanisms are lacking.[23] Of course, this cannot be true. All differentiation is active and requires multiple inputs and decision points. Furthermore, the so-called sex hormones are interconvertible. Both males and females secrete all of them and, for considerable parts of our lives, in not very different proportions. The historian Diana Long Hall has chronicled the discovery of these hormones and the ways gender ideology got incorporated into designating them "male" and "female."[24] And the biologist Anne Fausto-Sterling has noted that even the names scientists gave these hormones are ideology-laden. The so-called male hormone was named androgen, which is Greek for "generator of males," but there is no gynogen, no "generator of females." Instead the female analog of androgen is called estrogen, from *oestrus*, which means frenzy or gadfly in Greek.[25]

The standard description of fertilization follows the same traditional script.[26] Ejaculation launches sperm on their dauntless voyage up the female reproductive tract. By contrast, eggs are "released" or "shed" from the ovary to sit patiently in the fallopian tubes until a sperm "penetrates" and thereby "activates" them. Given that in fertilization two cells join together and their nuclei fuse, why is it that we say that a sperm *fertilizes* the egg, whereas eggs *are fertilized*? In 1948, the psychologist Ruth Herschberger caricatured this scenario in her delightful book *Adam's Rib*,[27] but that did not change the standard biological descriptions.

Of course, both sessile eggs and sprightly sperm are fabrications. Eggs, sperm, and the entire female reproductive tract participate in fertilization, and infertility can result from the malfunctioning of any of them. Also, a lot goes on in the egg's cytoplasm both before and after fertilization. As the biologist and historian of science Bonnie Spanier points out, modern biologists have focused excessively on the role of the sperm and on chromosomes and genes.[28] For this reason, they have paid much more attention to the fact that eggs and sperm contribute the same number of chromosomes and genes than to the fact that whereas the sperm contributes only its nucleus, the egg in addition contributes its entire cytoplasm to the embryo. This means that the embryo incorporates the entire cell substance and metabolic apparatus of the egg, and the entire egg, not just its DNA, plays a part in the differentiation and growth of the embryo.

If we turn at last to the surely objective realm of DNA molecules, we find a paper, published in December 1987 in the journal *Cell*, entitled "The Sex-Determining Region of the Human Y Chromosome. . . ."[29] Interestingly enough, this article was immediately publicized in the scientific weekly *Science* and in the daily press. The authors claimed, and magazines and newspapers promptly reported, that sex is determined by a single gene, located on the Y chromosome. The X chromosome, the authors wrote, has a similar gene, but according to them it has nothing to do with sex differentiation. Reading the paper more closely, we see that the authors identified a region on the Y chromosome that seemed to be correlated with the differentiation of testes. When this region was missing, no testes developed; when it was present, they did. By this argument, the presence or absence of testes determines sex: people who have testes are male; people who don't have testes are female.

More recently, in December 1989, two other groups of scientists have claimed that the region on the Y chromosome which the previous group had identified as "determining sex" does *not* "determine" either maleness or sex because males can develop testes in its absence,[30] but that has simply refocused the search for *the* sex gene elsewhere on the Y chromosome. None of these authors points out that being female implies more than not having testes, and that the differentiation of sex organs, whether female or male, requires processes in which many genes, proteins, and other metabolites must be involved.

These examples illustrate the ways in which our society's constructs of race and sex difference penetrate the biological sciences. When we use science to investigate subjects like race and sex, which are suffused with cultural meanings and embedded in power relationships, we need to be wary of scientific descriptions and interpretations that support, or even enhance, the prevailing political realities.

The same can be said about genes. DNA, the chemical, has material reality, but the concept of the gene, which long predates any thought of DNA, has been constructed to fill a host of political, economic, and cultural as well as scientific needs. For this reason, we would do well to become suspicious whenever characteristics are attributed to genes that neatly fit these rather inert molecules for their ideological tasks.

NOTES

1. J. B. S. Haldane, *New Paths in Genetics* (New York: Harper and Brothers, 1942), p. 11.

2. Londa Schiebinger, *The Mind Has No Sex? Women and the Origins of Modern Science* (Cambridge, Mass.: Harvard University Press, 1989), p. 165.

3. Allan Chase, *The Legacy of Malthus: The Social Costs of the New Scientific Racism* (New York: Alfred Knopf, 1977).

4. Stephen Jay Gould, *The Mismeasure of Man* (New York: W. W. Norton, 1981), p. 35.

5. Walter Rodney, *How Europe Underdeveloped Africa* (Dar-es-Salaam: Tanzania Publishing House, 1972), pp. 99–100.

6. Gould, *The Mismeasure of Man*, p. 103.

7. Cited in Dorothy Burnham, "Black Women as Producers and Reproducers for Profit," in Marian Lowe and Ruth Hubbard, eds., *Woman's Nature: Rationalizations of Inequality* (New York: Pergamon Press, 1983), p. 35.

8. Gunnar Myrdal, *An American Dilemma: The Negro Problem and Modern Democracy* (New York: Harper and Brothers, 1944), p. 1367.

9. Howard Zinn, *A People's History of the United States* (New York: Harper and Row, 1980), p. 406.

10. Leo Kuper, ed., *Race, Science and Society* (Paris: UNESCO Press, 1975).

11. R. C. Lewontin, Steven Rose, and Leon J. Kamin, *Not in Our Genes: Biology, Ideology, and Human Nature* (New York: Pantheon, 1984), esp. pp. 119–129.

12. Richard Lewontin, *Human Diversity* (New York: Scientific American Books, 1982).

13. Janet L. Norwood and Deborah P. Klein, "Developing Statistics to Meet Society's Needs," *Monthly Labor Review* 112; no. 10 (October 1989): 14–19.

14. Ibid.

15. Ibid.

16. Nancy Krieger and Mary Bassett, "The Health of Black Folk: Disease, Class, and Ideology in Science," *Monthly Review* 38 (July–August 1986): 74–85.

17. Mary T. Bassett and Nancy Krieger, "Social Class and Black-White Differences in Breast Cancer Survival," *American Journal of Public Health* 76 (1986): 1400–1403.

18. Kenneth C. Schoendorf, Carol J. R. Hougue, Joel C. Kleinman, and

Diane Rowley, "Mortality among Infants of Black as Compared with White College-Educated Parents," *New England Journal of Medicine* 326 (1992): 1522–1526. For an attempt to initiate this sort of global analysis, see Nancy Krieger, Diane Rowley, Allen A. Herman, Byllye Avery, and Mona T. Phillips, "Racism, Sexism, and Social Class: Implications for Studies of Health, Disease, and Well-Being," *American Journal of Preventative Medicine,* supp. to vol. 9, no. 6 (1993): 82–122.

19. Suzanne J. Kessler and Wendy McKenna, *Gender: An Ethno-methodological Approach* (Chicago: University of Chicago Press, 1978).

20. Edward O. Wilson, *Sociobiology: The Modern Synthesis* (Cambridge, Mass.: Harvard University Press, 1975).

21. Sarah Blaffer Hrdy, "Empathy, Polyandry, and the Myth of the Coy Female," in Ruth Bleier, ed., *Feminist Approaches to Science* (New York: Pergamon Press, 1986).

22. Ruth Hubbard, *The Politics of Women's Biology* (New Brunswick, N.J.: Rutgers University Press, 1990), pp. 87–118.

23. Anne Fausto-Sterling, *Myths of Gender: Biological Theories About Women and Men* (New York: Basic Books, 1985).

24. Diana Long Hall, "Biology, Sex Hormones, and Sexism in the 1920's," in Carol C. Gould and Marx W. Wartofsky, eds., *Women and Philosophy: Toward a Theory of Liberation* (New York: G. P. Putnam's Sons, 1976).

25. Anne Fausto-Sterling, "Society Writes Biology/Biology Constructs Gender," *Daedalus* 116 (1987): 61–76.

26. Emily Martin, "The Egg and the Sperm," *Signs* 16 (1991): 485–501.

27. Ruth Herschberger, *Adam's Rib* (New York: Harper and Row, 1948).

28. Bonnie Spanier, "Gender and Ideology in Science: A Study of Molecular Biology," unpublished manuscript.

29. David C. Page et al., "The Sex-Determining Region of the Human Y Chromosome Encodes a Finger Protein," *Cell* 51 (1987): 1091–1104.

30. M. S. Palmer et al., "Genetic Evidence that ZFY Is Not the Testis-Determining Factor," *Nature* 342 (1989): 937–939; Peter Koopman et al., "*Zfy* Gene Expression Patterns Are Not Compatible with a Primary Role in Mouse Sex Determination," *Nature* 342 (1989): 940–942.

Species, Races, and Genders:
Differences Are Not Deviations

David L. Hull

Science is a matter of mutual use. Scientists must use the work done by other scientists in order to further their own research, and reciprocally the greatest reward that one scientist can give another is to use his or her work, not just cite it but actually use it. A major objection to the Human Genome Project (HGP) as it was originally proposed is that relatively few scientists can see how they can possibly use the results of all this effort. They can think of much better ways to spend the huge amounts of money budgeted for the project, ways that would benefit their own research much more substantially and directly. Perhaps scientists may seem selfish for preferring research that they themselves can use. Instead they should be motivated primarily by the disinterested search of truth for its own sake, but such overly romantic notions have not proven to be very efficacious in other human endeavors. There is no reason to expect them to work any better in science. Harnessing individual good for the larger good, when it is possible, seems to be much the better strategy.

The most prevalent objection raised to the HGP is that raw sequence data alone are not very helpful. Once the HGP is complete, the most daunting and important task will remain: the elucidation of the functions of these sequences. As one author put this objection, "The world's most boring book will be the complete sequence of the human genome: three-thousand-million letters long, with no discernible plot, thousands of repeats of the same sentence, page after page of meaningless rambling, and the occasional nugget of sense—usually signifying nothing in particular. The manuscript should be finished by the year 2005, and the first copy will cost a great deal—about the same, in fact, as a Trident nuclear submarine."[1]

Because my assigned topic concerns the similarities and differences among species, races, and genders, I propose to discuss quite a different objection, one that arises in the context of evolutionary biology.[2] Species are the things that evolve, and subspecific variation of some sort is the raw material of evolution. At one time evolutionary biologists thought that something like races are incipient species—species in the making. Nowadays, this view is not widely held. Instead, much more localized intraspecific groupings are thought of as the precursors to full-fledged species. But at least races are one form of intraspecific variation, and intraspecific variation plays a central role in the evolutionary process. Gender is quite another matter. I will approach this issue from the perspective of sexual dimorphism. Many species are sexually dimorphic, and sexual dimorphism does play a role in the evolution of some species. The function of sex in the evolutionary process has been a major concern among evolutionary biologists over the past couple of decades. The role of sexual dimorphism is much less fundamental to our understanding of the evolutionary process but still worth discussing. I have much less to say about gender. Although the idea of gender is used primarily in connection with the human species, gender roles do exist in other species as well.[3] In any case, the question I address is, what bearing are the results of the HGP likely to have on these areas of research? My answer is, not much.

The main message of this essay is that, from the perspective of evolutionary biology, similarity and difference are of only secondary importance. Factors such as gene flow, geographic isolation, and population structure are of much greater importance. An additional message is that, in evolutionary biology, variation is not the same as deviation. We as human beings are strongly predisposed to view all variation as deviation. Ordinary people, including most scientists, view the mutations that are the raw material of evolution as not just different but deviant. They result from genes being "damaged." There are normal and abnormal genes, normal and abnormal characters, normal and abnormal individuals. As central as this notion of deviation is to many areas of human inquiry, it has no role in evolutionary biology.

SPECIES

Although, from the outside, evolutionary theory seems easy enough to understand, it is anything but. People who would be somewhat reluctant to voice their opinions on quantum theory feel distressingly free to share their feelings about evolutionary theory. But evolutionary theory is extremely counterintuitive, and one source of the difficulty that we have in understanding the evolutionary process turns on intra- and interspecific variability. What people find difficult to grasp is that the essence of species is variation, both at any one time and through time. Any species that is not internally genetically heterogeneous is likely to go extinct quite rapidly. Without such variation, evolution would grind to a halt. On current estimates, however, we have nothing to worry about. Cheetahs are few in number and genetically quite homogeneous. We human beings are anything but an endangered species. We have multiplied far beyond the biblical injunction. We are also genetically quite heterogeneous.

Most people, to the contrary, are convinced that every species can be characterized in terms of a single, tightly clustered set of characters. They assume that the vast majority of organisms that belong to a species are "normal" or "typical," a few deviants notwithstanding. They are willing to admit that variation occurs but insist on interpreting it as deviation from a norm. Ordinary people (and in this context this designation includes everyone except evolutionary biologists) are convinced that there are typical crows, typical oak trees, and, most important, typical human beings. They are convinced that a set of characters can be enumerated that all, or almost all, organisms that belong to a species possess. To put the matter more formally, particular species have essences. Perhaps these essences change, but this change takes long stretches of time. A character that used to be universally distributed in a species can become modified into another character or replaced by something else, but this process takes hundreds of thousands of years. At any one time, so most people are convinced, each species must have its own unique essence.

In population genetics, the preceding conviction was encapsulated in the term "wild type." At any locus, one allele was thought to prevail—the wild type. All other alleles are departures from the wild type and are either rapidly eliminated because they are less fit or rapidly spread through the population if they are more fit. It turns out that those alleles commonly designated as the wild type tended to have only one thing in common. They characterized those organisms that happened to live closest to the nearest available road. Different sorts of variation can occur at a particular locus. Some cases are like eye color in human beings. Slightly fewer than 1 percent of human beings have blue eyes. Since the allele for blue eyes is recessive to the allele for brown eyes, probably 2 or 3 percent of people carry this "damaged" gene. In other cases, a couple of alleles are prevalent and a few others rare, for example, the ABO blood group. In other cases, no allele even approaches 50 percent.

Once the right techniques became available, we discovered that we had seriously underestimated how variable species are. At the molecular level more variation exists than is reflected at the phenotypic level. Much of this variation turns out to be adaptively neutral. No matter which allele is present, the result is the same. We then discovered that most of the genetic material has no discernible function at all. It doesn't do anything but replicate. Such junk DNA is of no use to the organism, although some of it may have served a function at some previous time and might become functional again in the future. As useless as this junk DNA is to the organism, it turns out to be very useful for biologists interested in inferring the phylogenetic development of species. Junk DNA is just as good at providing clues to descent as functional DNA, and in several respects better.

When one turns one's attention from intraspecific variation to interspecific variation, the situation becomes even more counterintuitive. It is bad enough that so much variation occurs within particular species, but we could handle it if reasonably sharp gaps separated such amorphous clusters. Such is not the case. From the perspective of character covariation, species frequently blend imperceptibly into one another. A few decades back, an extremely bright and energetic group of systematists decided that the goal of systematics should be to classify organisms according to varying degrees of overall similarity.[4] They identified dozens of character states, fed them into a computer, and then, using various algorithms, produced groupings. After some initial progress, impregnable barriers were confronted. No two programs gave the same results. In some cases, a slight variation in the number of species studied had massive effects; in others not. Using other programs, additional characters could be added without making much of a difference; in others, new characters produced radical changes in the classifications. The notion of overall similarity turned out to be one more will-o'-the-wisp. To make matters worse, when organisms are clustered solely in terms of varying degrees of similarity, the results are totally unacceptable, for example, placing males and females in separate species.

For organisms that reproduce sexually, a criterion exists for distinguishing species that is more fundamental than degrees of similarity—mating. Of course, this criterion does not always produce absolutely sharp boundaries between species. In many species, hybrid zones exist. When certain species meet, some of their members mate with some degree of success, but these hybrid zones remain narrow and the two species do not merge into one. At times, however, interspecific introgression can be quite substantial. More commonly, species remain subdivided into geographically isolated populations for long periods of time without speciation occurring. But at the very least, gene flow provides an additional criterion to gene and character distribution for distinguishing species.

Numerous difficulties remain in deciding which groupings of organisms count as species and even whether or not all organisms form species. After all, for the first half of life on Earth, nothing like sexual reproduction occurred, certainly not meiosis. Many organisms today still do not reproduce sexually. Species are the things that evolve, but the question remains whether or not a single level of organization exists across all organisms that deserves to be singled out for special attention. Will the HGP aid in solving these problems? Not that I can tell. The HGP is not designed to study intra- or interspecific variation. To be sure, a few dozen different genomes are being used in the HGP, but such sampling is not the sort that population geneticists need. Walter Bodmer has announced a program to study human genetic diversity, but if this project ever materializes, it will be an *addition* to the HGP as it is now constituted.[5]

The HGP is limited to the study of the human species. As it turns out, the human species is not all that different genetically from chimpanzees and gorillas, but at the phenotypic level, there is no problem. In addition, the human species is reproductively isolated from all other species. No hybrids are possible. Given these sharp phenotypic and reproductive gaps, the lack of sharp distinctions at the genetic level is of little consequence. Fortunately, we do not need the sort of data that the HGP will generate to distinguish the human species from other species, because it would not help very much anyway. To distinguish species, we need information about such things as geographic variation and interfertility, and the HGP is not designed to produce such information. The results of the HGP, however, will add a few data points for biologists interested in phylogeny reconstruction.

RACES

In a reasonably high percentage of cases, species are sharply distinct from each other, whether or not human observers can tell the difference. In other cases, speciation is not complete or is in the process of breaking down. The sort of variation that occurs below the species level is almost always of the more problematic sort. Neither character nor physiological gaps are sharp. In connection with the human species, one commonly hears that races exist only in the mind of the racist. Maybe yes; maybe no. Even though infraspecific variation is even more problematic than variation between species, systematists feel the need to subdivide species into less inclusive groups such as varieties, subspecies, and races. Some systematists are interested in subdividing species no matter how arbitrary the criterion. For example, there is something in systematics termed the 75 percent rule. According to this rule, subspecies are recognized if 75 percent of the specimens under investigation can be placed unequivocally in one subspecies or another; otherwise not.

However, most systematists are interested in discerning subspecific groupings that play a role in the evolutionary process. All of the preceding discussion concerns variation across entire species, but the variation that is relevant to speciation is geographically localized. For example, as far as the ABO blood group is concerned, A and O are the most common alleles, but their prevalence varies geographically. Blood type A is common among Eskimos, Europeans, Japanese, and in some areas of Australia. It is extremely rare in the indigenous populations of South America and Alaska. Type B is extremely widespread in Asia and central Africa, and all but absent in the Western Hemisphere. Type O has just the opposite distribution. If one looks at geographic distributions rather than at just abstract variation, one can discern subspecific groupings. Among these groups, some boundaries are fairly sharp; for example, Australian aborigines, Basques, and Navaho Indians form genetically discrete groups. However, most of these infraspecific subdivisions are anything but sharp and are getting even less sharp as migration allows for increasingly panmictic mating.

The question is—what sorts of variation contribute to one species splitting into two or more species? At one time, evolutionary biologists thought that a single, widely distributed species commonly split into two or more subspecies (or races) and that these subspecies gradually diverge until they become species in their own right. The current view among evolutionary biologists is, however, that this mode of speciation is not especially common. New species arise from much smaller groups of organisms—single geographic isolates. As long as evolutionary biologists thought that a half-dozen or so widely distributed parts of a species are the sources for new species, all the work necessary to delineate them seemed worthwhile. Is a particular species made up of five subspecies or twelve? Now that evolutionary biologists think that such widespread variation is merely an effect of processes occurring at much lower levels of analysis, the subspecies question has ceased to be a hot topic for all species but one—ours. At one time the attention paid to races of human beings had some biological foundation. Now it has very little. Races are of social interest only, and socially recognized races have next to nothing to do with the distinctions that biologists make.

In short, our everyday experience leads us to believe that, by and large, we can tell species apart just by looking. Organisms that look alike are alike, some accidental variations notwithstanding. With considerably less justification, we think that we can do the same with respect to subspecies. Our habit of thinking stereotypically has developed a very bad reputation. However, stereotypical thinking is a very useful heuristic when we don't have the time, knowledge, or ability needed to engage in the sort of activity necessary for more accurate decisions. In most contexts, we don't have to be right all the time. Good enough has to be good enough. But evolutionary biologists cannot be content with such rough and ready characterizations. The distinctions that common sense blurs, evolutionary biologists must keep in the forefront of their attention.

How will the HGP help evolutionary biologists, in particular those concerned with population genetics? Genes are certainly central to the evolutionary process, so central that some evolutionary biologists argue that we can understand the evolutionary process ignoring everything else about it, including organisms. Evolution is nothing but changes in gene frequencies. Gene selectionism to one side, evolutionary biologists would like to know the genetic makeup of the organisms that they are studying. Unlike developmental biologists, they are interested in all genes, not just the minority of genes that are now functional. But they are interested in the *distribution* of the various alleles of a gene. Just being told that a particular stretch of DNA is characterized by a particular sequence of bases tells population geneticists almost nothing of interest. They need to know which alleles exist and how they are distributed. When it comes to questions of adaptations, they would also like to know how they function. The HGP provides little help here as well.

I do not know about people in general, but I for one am getting a little tired of all the attention that is being devoted to language at the expense of the phenomena described by language. Considerable effort, for example, was expended to decide what to call the virus that causes AIDS. After much deliberation, a committee picked a name whose acronym begs to be misused as we are treated to references to the HIV virus, as propitious an expression as "ATM machines," or the "HGP project" for that matter. Pedantry to one side, the language that we use can get us into trouble. For example, I wonder how much our vertebrate biases influenced early efforts to figure out the structure of DNA. Initially everyone working on the project attempted to construct a molecule with the backbone on the inside—where it belongs. The big conceptual shift made by Watson and Crick was, after all, to put the backbone on the outside, to shift from a vertebrate to an invertebrate model.

Even such harmless-looking words as the definite article can get us into trouble. We commonly hear about *the* duck, or *the* mouse, or even *the* human being. In a recent article, John Maddox set out the case for the HGP. His entire discussion was predicated on the assumption that something exists that can be properly termed the "complete sequence of a presumably representative human genome."[6] People seem to think that they know what they are talking about when they refer to *the* human being or *the* human genome. I don't think so. What blood type at the ABO locus does *the* human being possess? What eye color, hair texture, skin color, body height, body weight, sex? So, the story goes, the famous systematist Carolus Linnaeus has been "honored" posthumously by being designated as the type specimen for *Homo sapiens*. Hence, a relatively short Swede is in some significant sense the "typical" human being, but the vast majority of human beings have brown eyes, black hair, and dark skin.

One might agree that Linnaeus is not a very good choice for a type specimen for the human species but insist that some other human being would be. However, the more one learns about the evolutionary process, the more suspicious one becomes of the very notion of a "typical" member of any species. The essence of particular species is to have no essence. Any attempt to characterize the genetic makeup of any species, including the human species, in terms of typical, normal, or wild-type alleles and departures from these norms as "deviations," introduces falsification of such magnitude as to make the evolutionary process incomprehensible.

The major focus of this conference is supposed to be the HGP. Although there may be something properly termed *the* HGP, there is no such thing as *the* human genome. But advocates of the HGP might complain that such references, though technically inappropriate, are really harmless enough. For example, a recent issue of *Science* included a wall chart which lists the current results of the mapping and sequencing of "the fruit fly," when several thousand species of fruit fly exist, and later labels a drawing as the "wild-type female" for *Drosophila melanogaster*. What harm can such references do? Sometimes specialists do get overly picky. For example, although the objections recently raised by several evolutionary biologists to the use of DNA fingerprinting to identify particular people are technically legitimate, they assume standards of evidential certainty foreign to both science and the law. However, the complaints that evolutionary biologists are raising to the stereotypical thinking implicit in the HGP are not mere pedantry. They rest on the very foundations of their science. Once the HGP is complete, and the hundreds, possibly thousands, of sequencing slaves have been unchained from their laboratory benches, evolutionary biologists will still not know what they need to know for their activities. They will have a complete description of a few collated human genomes, that is all, and such a description will not do them much good.

GENDER

So far I have talked only about species and subspecies (or races). How about gender? I don't know much about gender. One thing that I do know is that it is extremely dangerous to discuss this complex of issues, especially when one does not know very much about it. I do know something about sex. For roughly the first half of life on earth, there was some gene exchange but no reduction division. No dance of the chromosomes, no fertilization, no cost of meiosis. But once such machinery evolved, it spread quite rapidly. A lot of organisms still do not reproduce sexually, but quite a few do. Creationists are fond of raising objections to present-day evolutionary theory of the sort that Darwin's contemporaries raised in the nineteenth century. Sooner or later they are bound to find out about the scandal posed by the cost of meiosis. How can something that seems to have a 50 percent cost evolve in the first place, let alone proliferate? Among organisms that do engage in sex, how come the usual sex ratio is one to one? What special roles do more variable sex ratios play? These are the sorts of questions that evolutionary biologists ask about sex, and I cannot see that the HGP will contribute much toward answering them.

In any case, sex did evolve, and in many species the gametes can easily be distinguished. Some are relatively large and sessile; others small and motile. The former are female; the latter male. Gametic fusion can take place without such dimorphism, but in many species, including our own, it exists. Gametes can be dimorphic without organisms exhibiting the same degree of sexual dimorphism. In many species, the sexes are indistinguishable, not just to us but to the organisms themselves. They engage in sex and get it right half the time. But in some species there are noticeable differences. The sexes are strongly dimorphic, and these differences play a significant role in the mating process.

Sexual differences are determined in a myriad of different ways. Sometimes any zygote in a species can become either male or female, depending on the environment. In other cases, which sex a zygote turns out to be is almost totally a function of its genetic makeup, the environment providing a largely indeterminant background. The human species is nearer the latter pole. Most people are either XX or XY. Of these, most individuals with XX exhibit a female phenotype, and most with XY exhibit a male phenotype. As the organizers of Olympic competitions have discovered to their dismay, a few individuals with two XX chromosomes can be phenotypically male, and a few XY individuals can be phenotypically female. The fact that reference to "normal" males and "normal" females seems to make sense in such contexts is a good sign that we have left the realm of evolutionary biology and are addressing questions about development. Within the context of development, certain variations can count with some justification as being "deviations." However, even in development we are all too quick to label every variation a deviation. Nowadays we are unlikely to characterize females as deviant males, but there is a strong tendency to characterize "masculine" women and "feminine" men as deviations from some socially charged norm.

Some issues in science are relatively straightforward. If people do not understand them, they are simply not trying. However, other issues are all but impossible to explain in English or any other natural language. The nature-nurture issue is of this second sort. All the goodwill in the world is necessary to understand what is going on, and if this area of discussion is characterized by anything, it is not a surfeit of goodwill. People want desperately not to understand. For this reason and others, I do not intend to let myself get drawn into that dreary bog. The question before us is, however, what implications does the HGP have on the issue of gender differences in human beings? A specification of "the" human genome will no more contribute anything of significance to answering these questions than to any others of the same sort. What is needed is a study of the effects of allelic differences on the phenome in a variety of environments.

In several species of organisms, particular loci have been discovered at which allelic differences influence the determination of sex.[7] Will similar genes be discovered in human beings? Possibly. How about gender differences? Possibly, but perhaps not. Recently, a researcher has claimed to discover a few cells in the hypothalamus that differ systematically between heterosexual males, on the one hand, and heterosexual females and homosexual males, on the other hand. Homosexual females were not included in the study.[8] Such studies engender considerable controversy, as if they were socially relevant. I find it difficult to join in all the enthusiasm. Regardless of the contributions of genes and environment in determining whether a human being is male or female, masculine or feminine, homosexual or heterosexual, my attitudes toward myself and others remain unchanged.

In this respect I am a Martian. Most people think that very important issues such as social, moral, and legal responsibility hinge on the outcomes of these empirical investigations. Perhaps they should, but I am afraid they will not. Some people think that sex preference is strongly influenced by one's genes, others think that the first few years of rearing do the trick, still others think that it is a combination of the two. There are even people who think that teenagers actually *choose* which preference to have. But discomfort, hate, and persecution flourish regardless of empirical data. No one thinks that we can choose our sex or race, but sexism and racial prejudice persist. We can choose which religion to join. Religious prejudice persists. In matters such as these, genes, environment, and even choice appear to be only selectively relevant.

What implications does the HGP have for issues concerning human responsibility? None that I can see, but then I hope advocates of this project do not think there are any. It is really not fair to criticize a program for not doing what no one ever intended it to do. The great fear is that knowledge of particular human genomes will raise all sorts of psychosocial and legal problems. In principle it may. In practice, it is unlikely to raise very many problems very often. Our greatest protection in this regard is the cost involved. When only a few dollars are spent per year on the medical care of most people in the world, it does not seem likely that the significant financial expenditure necessary to produce a map of the genomes of very many people will be forthcoming. The vast majority of people in the world are safe. Only the relatively few people living in rich, industrialized countries need worry. From the literature on this topic, it seems that we are the only ones who count anyway.

In sum, the objection that I have raised throughout the body of this essay is that the HGP does not do enough. It provides only a specification of an archetypical human genome, a specification that omits all the variability that is so important to evolutionary biologists and the developmental sequences so central to developmental biology, but both deficiencies can be remedied. Then the fear arises that scientists are in the position to do too much—to specify the genomes of particular human beings. Such knowledge, like all knowledge, can be abused. Perhaps I am being too optimistic, but when we lack adequate funds in Illinois to vaccinate all our children against measles, I suspect that we will not have sufficient money to sequence very many of their genomes. Perhaps some yuppies may have both the money and the inclination to fix up their fetuses so that they get the children that they deserve, but I doubt that this practice will become widespread.

NOTES

1. J. S. Jones, "Songs in the Key of Life," *Nature* 354 (November 1991): 323.

2. S. Sarkar and A. I. Tauber, "Fallacious Claims for HGP," *Science* 353 (October 1991): 691.

3. For a history of sexual selection, see Helena Cronin, *The Ant and the Peacock* (Cambridge: Cambridge University Press, 1991).

4. P. H. A. Sneath and R. R. Sokal, *Numerical Taxonomy* (San Francisco: W. H. Freeman, 1973).

5. Walter Bodmer, "Human Genome Dispute," letter to editor, *Nature* 354 (December 1991): 426.

6. John Maddox, "The Case for the Human Genome," *Nature* 352 (July 1991): 11–14.

7. A. McLaren, "What Makes a Man a Man?" *Nature* 346 (July 1990): 216–217.

8. S. LeVay, "A Difference in Hypothalamic Structure between Heterosexual and Homosexual Men," *Science* 253 (1991): 1934–1937.

The Ethics of Human Germ-Line Genetic Intervention

LeRoy Walters

The Human Genome Project (HGP) will make possible new technical possibilities that fall into two large categories: improved diagnostic capabilities and improved capabilities to modify human cells, tissues, organs, and even organisms by genetic means. These latter capabilities are often called "therapeutic," but as we shall see, the word therapeutic may be too narrow to encompass some future technical possibilities. This essay will focus on the new possibilities of human genetic modification that the HGP may render technically feasible.

Within the sphere of human genetic modification, one can draw conceptual distinctions among four possible types of intervention as seen in figure 1:[1] The vertical line in this matrix distinguishes between genetic changes that affect only somatic (or nonreproductive) cells and those that affect germ-line (or reproductive) cells. This distinction is important, for it marks the difference between genetic changes that involve only the person undergoing an intervention (Types 1 and 3) and genetic changes that will be passed on to the descendants of the person undergoing the intervention (Types 2 and 4). The horizontal line distinguishes between interventions that treat or prevent disease (Types 1 and 2), on the one hand, and those that improve on the capabilities of the average healthy human being (Types 3 and 4), on the other. There are cases that seem to straddle the horizontal line, for example, improving the ability of our immune systems to fight disease or decreasing our forgetfulness, but these borderline cases are beyond the scope of this essay.

FIGURE I. FOUR TYPES OF POTENTIAL GENETIC
INTERVENTION

	Somatic	Germ-line
Cure or Prevention of Disease	1	2
Enhancement of Capabilities	3	4

I will focus on Type 2 genetic intervention in this essay—that is, on germ-line genetic intervention for the cure or prevention of disease. Further, so that we may avoid the problem of borderline cases, I will stipulate that the diseases to be treated or prevented must be acknowledged by all to be *serious* diseases. Examples of such diseases are cystic fibrosis, Huntington disease, sickle cell anemia, Tay-Sachs disease, and Duchenne muscular dystrophy. I should also note that in examining the ethical issues surrounding human germ-line intervention, we will be looking into the future. While such intervention is technically feasible now, and is routinely performed with several species of laboratory animals, it is not currently being proposed for human use because available techniques are so unpredictable in their outcomes.

THE MAJOR NEEDED TECHNOLOGICAL BREAKTHROUGH: GENE REPAIR OR GENE REPLACEMENT

There are currently about ten active somatic-cell gene therapy protocols in the world.[2] These are examples of Type 1 genetic intervention. All of the active studies employ a technique that might be called gene addition. For readers who use word-processing software, I can perhaps best explain gene addition by saying that it is very much like using the "insert" command. Consider the graphic representation in figure 2.

In this illustration genes are represented by words, and we are looking at a single strand of DNA. The word "Wiao" is a misspelled word, a mutation. If this mutation occurs in a critical point in the genome of an individual, it can affect the production of something that is essen-

FIGURE 2. GENE ADDITION

Malfunctioning Gene and Surrounding Genes
   —The-University-of-*Wiao*-in-Iowa-City—

Command: Insert the Properly Functioning Gene, Surrounding Genes, and Vector

   —[Vector]>-The-University-of-*Iowa*-in-Iowa-City—

The added gene integrates into a random site in a random chromosome. There it begins to function within the cell.

tial for the body's functioning. For example, in the disease called severe combined immune deficiency, which affects some children, the malfunctioning of a single gene that normally produces a single enzyme can block the functioning of the entire immune system.

In current studies multiple copies of a vector or vehicle are hooked to multiple copies of a properly functioning gene that has been derived from a normal human cell. The vector-gene combinations are delivered into the nuclei of multiple somatic cells, for example, white blood cells or liver cells. The usual vector is a domesticated, carefully engineered retrovirus that in its unmodified form infects mice. Unfortunately, this kind of vector is really an unguided missile. That is, in any given cell one cannot predict into which chromosome the vector will deliver the new gene or at what point on the chromosome it will insert the gene.

Gene addition has worked fine for somatic-cell gene therapy, especially in the Blaese-Anderson study that is treating children with severe combined immune deficiency.[3] As long as one normal copy of the gene is present in a somatic cell and making its product, the cell should be able to function properly. However, gene addition would not seem appropriate for germ-line intervention. If researchers simply *added* properly functioning genes to malfunctioning genes in reproductive cells, both types of genes—the normal and the abnormal— would be sent forward into future generations.

What is needed, therefore, for effective germ-line intervention is a major technological breakthrough, namely, a technique for *replacing* malfunctioning genes with properly functioning genes in reproductive cells.[4] Such precise targeting of genes may not be feasible for a very

FIGURE 3. GENE REPAIR OR GENE REPLACEMENT

Malfunctioning Gene and Surrounding Genes
  —The-University-of-*Wiao*-in-Iowa-City—

Command: Search for the Malfunctioning Gene and Surrounding
Genes
  —The-University-of-*Wiao*-in-Iowa-City—

Command: Delete the Malfunctioning Gene and Surrounding Genes,
and Replace Them with the Properly Functioning Gene and
Surrounding Genes

—[Vector]>-The-University-of-*Iowa*-in-Iowa-City—

The malfunctioning gene is "snipped out," and the properly
functioning gene is "spliced in" at precisely the same location on the
same chromosome in the cell. The gene then begins to function.

long time. Using the word-processing analogy, let me illustrate two
possible strategies for gene replacement; see figure 3.

In this first case, the exact "spelling" of the mutation is known, and
the vector is programmed to look for a particular combination of letters, to remove that combination from the text, and to replace the combination with the properly spelled word. In this illustration, the misspelled word "Wiao" is removed and replaced by the properly spelled
word "Iowa."

A second strategy for gene replacement is illustrated in figure 4.

In this second case the precise spelling of the mutation is not
known. In fact, there may be several different misspellings, all of
which cause a genetic disease, as in the case of cystic fibrosis. However, the context in which the misspelled word occurs is known. In
this case someone using a word-processing program would undoubtedly use wild-card characters to say the following: "After the words
'The-University-of-' and before the words '-in-Iowa-City' there occurs
a misspelled word. I do not know how many letters there are in the
misspelled word, so I will simply represent the word by four asterisks.
[In WordPerfect 5.1, a single asterisk would function just as well to
represent an unknown number of intervening characters.] Whatever
the word is that falls between the two phrases noted above, delete that
word and replace it with the word 'Iowa'."

FIGURE 4. GENE REPAIR OR GENE REPLACEMENT

Malfunctioning Gene and Surrounding Genes
  —The-University-of-@&#!-in-Iowa-City—

Command: Search for the Malfunctioning Gene and Surrounding Genes
  —The-University-of-****-in-Iowa-City—

Command: Delete the Malfunctioning Gene and Surrounding Genes, and Replace Them with the Properly Functioning Gene and Surrounding Genes

—[Vector]>-The-University-of-*Iowa*-in-Iowa-City—

The malfunctioning gene is "snipped out," and the properly functioning gene is "spliced in" at precisely the same location on the same chromosome in the cell. The gene then begins to function.

SITUATIONS IN WHICH GERM-LINE INTERVENTION MAY BE PROPOSED

The questions arise: Why should anyone be thinking about or proposing Type 2 genetic intervention? What would be the rationale for this innovation? The answer to these questions is based primarily on the inherent limitations of the somatic-cell approach to gene therapy. Here is a first possible scenario.

SCENARIO I
Both the wife and the husband are afflicted with a recessive genetic disorder. That is, both have two copies of the same malfunctioning gene at a particular site in their chromosomes. Therefore, all of their offspring are likely to be affected with the same genetic disorder.

This kind of situation is likely to arise as medical care succeeds in prolonging the lives of people with genetic disorders like sickle cell disease or cystic fibrosis. In addition, scenario 1 may become more prevalent precisely if somatic-cell gene therapy is employed with large numbers of people afflicted with recessive genetic diseases. When these people are treated successfully by means of somatic-cell gene therapy, their diseases will be ameliorated and some of their somatic cells will be enabled to function normally, but their reproductive cells will remain the same. Even though they have been successfully treated, they will be carriers of genetic disease to the next generation. In fact, because all of their somatic cells contain two copies of a malfunctioning gene, it seems likely that, after meiosis, all of their sperm or egg cells will carry one copy of the malfunctioning gene. If two such phenotypically cured people marry and have children, all or almost all of their children will be afflicted with the disease with which their parents had once been afflicted. These children will then need somatic-cell gene therapy for the treatment of their disease, and so on through the generations.

There is a second scenario, in which multiple outcomes of reproduction are possible.

SCENARIO 2
Both the wife and the husband are carriers of a recessive genetic disorder. That is, each has one copy of a properly functioning gene and one copy of a malfunctioning gene at a particular site in their chromosomes. Following Mendel's laws, 25 percent of the couple's offspring are likely to be "normal," 50 percent are likely to be carriers like their parents, and 25 percent are likely to be afflicted with the genetic disorder.

This second kind of scenario is frequently encountered by genetic counselors. In this case, germ-line genetic intervention could be viewed as an alternative to prenatal diagnosis and selective abortion of affected fetuses or to preimplantation diagnosis and the selective discard of affected early embryos. In addition, a couple might elect germ-line genetic intervention in order to avoid producing children who are carriers of genetic defects, even if the children are not themselves afflicted with genetic disease. The parents would know that children who are carriers may one day face precisely the same kind of difficult reproductive decisions that they as parents are facing.

Here is a third scenario.

SCENARIO 3
One spouse carries the gene for a dominant genetic disorder that begins to afflict those who carry the gene at age forty. Each child of the couple has a 50 percent chance of inheriting the dominant gene and of being afflicted with the disorder at age forty.

This third scenario illustrates a serious limitation of gene addition, the technique currently employed for somatic-cell gene therapy. A dominant gene is by definition one that is capable of overwhelming the presumably normal gene that is present at the corresponding site in the affected person's cells. Thus, adding more copies of the normal gene is not likely to turn off the destructive dominant gene or overcome its deleterious effects. Only the removal of the dominant gene and its replacement by a properly functioning gene is likely to be successful in preventing disease.

In a fourth scenario temporal factors in early human development are predominant.

SCENARIO 4

A genetic disorder results in major, irreversible damage to the brains of affected fetuses during the first trimester of pregnancy. There is no known method for making genetic repairs in the uterus during pregnancy. Therefore, if any genetic repair is to be made, it must be completed before the embryo begins its intrauterine development.

In this case, somatic-cell gene therapy *might* be effective if one could deliver it to the developing embryo and fetus during the earliest stages of pregnancy, that is, shortly after the embryo has implanted in the uterus. However, there is no known method of administering intrauterine therapy to an early embryo, and a deferral of treatment until the second or third trimester, when some current fetal treatments can be administered, would allow irreversible damage to occur. Thus, it would appear that preimplantation treatment, which would almost certainly affect the future germ-line cells as well as the future somatic cells, may be the only feasible approach—especially for couples who reject the alternative of selectively discarding early embryos.

WHEN AND HOW GERM-LINE INTERVENTION COULD BE PERFORMED

There are three stages at which new genetic information could be introduced into the human germ line. The first and perhaps least technically complicated stage is the stage of preimplantation development by the early human embryo. The earliest point at which new genes could be introduced would be during the fertilization process itself, after the sperm has penetrated the surface of the egg but before the pronucleus of the sperm has fused with the pronucleus of the egg. Alternatively, new genes could be introduced with the aid of a vector or by microinjection when the preimplantation embryo had developed to what is called the blastocyst stage, perhaps six or seven days after fertilization. A third possibility would be to remove one or more totipotent cells from the four-, eight-, or sixteen-celled early preimplantation embryo and to make genetic modifications in the removed cells. At all of these stages new genetic material has been introduced into non-human preimplantation embryos, but often with high embryo loss rates and with unpredictable results in terms of the number of genes delivered to the cells and the sites at which the genes were incorporated into the genomes of the cells.[5]

A second stage at which germ-line genetic intervention could, at least in principle, occur is in a test tube or Petri dish with sperm or egg cells that are being prepared for in vitro fertilization.[6] In the case of sperm cells, it seems likely that the genetic change would need to be introduced into large numbers of cells in a predictable and repeatable way. In the case of egg cells, the numbers would be smaller, but the predictability and repeatability of the procedure would need to be equally assured.

A third possible stage for germ-line intervention would be in the reproductive organs of males or females. Here one must envision a vector that is not only capable of performing precise gene replacement or gene repair but also a vector that is capable of targeting the reproductive organs after having been introduced into another part of a person's body—for example, into a vein in the person's arm. In the case of males, the genetic modification would have to occur in the sperm-producing cells because the sperm present in the body at any given time will have been replaced within a few months. In the case of females, the genetic modification would be targeted to the preovulatory eggs that have been present in the woman's body since the latter stages of her own gestation.

MORAL ARGUMENTS FOR GERM-LINE INTERVENTION [7]

A first argument in favor of germ-line intervention is that it may be the only way to prevent damage to particular biological individuals when that damage is caused by certain kinds of genetic defects. One thinks, for example, of scenario 4 above, and of genetic disorders that may adversely affect the brains of embryos or fetuses in the early stages of pregnancy. Here the primary intent of gene therapy would, or at least could, be to provide gene therapy for the early embryo. An unintended side effect of the intervention would be that all of the embryonic cells, including the reproductive cells that would later develop, would be genetically modified. As noted earlier, the alternative to the preimplantation repair of embryos in this case would be selective discard of genetically impaired preimplantation embryos.

A second moral argument for germ-line genetic intervention might be advanced by parents. It is that they wish to spare their children and grandchildren from either (a) having to undergo somatic-cell gene therapy if they are born affected with a genetic defect, or (b) having to face difficult decisions about passing on a known genetic defect to their own children and grandchildren. This justification is at least akin to the justification that parents give for submitting their children to immunizations against polio, measles, and hepatitis B.

A third moral argument for germ-line intervention is more likely to be made by health professionals, public health officials, and legislators casting a wary eye toward the expenditures for health care. This argument is that, from a social and economic point of view, germ-line intervention is more efficient than repeating somatic-cell gene therapy generation after generation. From a medical and public health point of view, germ-line intervention fits better with the increasingly preferred model of disease prevention and health promotion. In the very long run, germ-line intervention, if applied to both affected individuals and asymptomatic carriers of serious genetic defects, would have a beneficial effect on the human gene pool and the frequency of genetic disease. (For further discussion of these and other moral arguments see the December 1991 issue of the *Journal of Medicine and Philosophy*.)[8]

MORAL ARGUMENTS AGAINST GERM-LINE INTERVENTION [9]

A first argument against germ-line intervention is that if the technique has unanticipated negative effects, those effects will be visited not only on the recipient of the intervention himself or herself but also on all of the descendants of that recipient. This argument seems to assume that a mistake, once made, could not be corrected, or at least that the mistake might not become apparent until the recipient became the biological parent of at least one child. For that first child, at least, the negative effects could be serious, as well as uncorrectable.

A second argument against germ-line intervention is that the technique would give human beings, or a small group of human beings, too much control over the future evolution of the human race. This argument does not necessarily attribute malevolent intentions to those who have the training that would allow them to employ the technique. It implies that there are built-in limits that humans ought not to exceed, perhaps for theological reasons, and at least hints that corruptibility is an ever-present possibility for the very powerful.

Third, one could argue that germ-line intervention would probably be used by dictators to produce a master race or mighty armies or submissive subjects. Here the pessimism about the corruptibility of human nature, or at least of some humans, is much more explicit. According to this view, Aldous Huxley's *Brave New World* should be updated, for modern molecular biology provides tyrants with tools for modifying human beings that Huxley could not have imagined in 1932.

A BRIEF EVALUATION OF THE ETHICAL ARGUMENTS

If there are plausible replies to the moral arguments against germ-line intervention, then there should remain at least a prima facie case for the technique. I will attempt to respond to the three objections in reverse order. The third objection, which cites the possibility that germ-line intervention will be deliberately misused by tyrants, may focus too much attention on technology and too little on politics. There is no doubt that bona fide tyrants have existed in the twentieth century and that they have made use of all manner of technologies—whether the low-tech methods of surgical sterilization or the annihilation of concentration camp inmates with poison gas, or high-tech weapons like nuclear warheads and long-range missiles—to terrify and to dominate. However, the best approach to preventing the misuse of genetic technologies may not be to discourage the development of the technologies but rather to preserve and encourage democratic institutions that can serve as antidotes to tyranny. A second possible reply to the tyrannical misuse objection is that germ-line intervention requires a long lead time, in order to allow the offspring produced to grow to adulthood. Tyrants are often impatient people and are likely to prefer the more instantaneous methods of propaganda, intimidation, and annihilation of enemies to the relatively slow pace of germ-line modification.

The second objection is a composite. It argues against entrusting germ-line intervention to a small group of experts who might thereby achieve uncontrolled or even uncontrollable power. To this objection, one can reply that so long as the technical experts are held publicly accountable and are asked to accept socially agreed-upon limitations on the use of their expertise, they will not be exercising uncontrolled power. On a more modest scale, the pioneers of somatic-cell gene therapy have been willing to subject their work to intense public scrutiny and rigorous prior review before proceeding to enroll human subjects in their studies. Nothing less would be expected if and when molecular biologists develop safe and effective techniques for performing germ-line intervention in humans. The other facet of this objection is a concern that human beings will somehow exceed inherent limits, perhaps even attempt to "play God," if they deliberately intervene in the course of human evolution. One reply to this objection is that certain therapies—for example, radiation and some chemotherapies—and the practice of medicine in general already produce major evolutionary effects. These effects are diffuse, to be sure, and are often unintended. A more ambitious reply is based on a more activistic view of the role of human beings in the world. This view is not satisfied to see genetic diseases being transmitted from one generation to the next without a vigorous effort to intervene. Rather, provided that the free choices of human beings are respected, this perspective is willing to employ all reasonable means to reduce the incidence of disease in future generations, including germ-line intervention.

Finally, it is true that mistakes may be made in the early trials of germ-line intervention and that some human beings will be injured as a result of this research. This possibility will be minimized if there is an interinstitutional, broadly based public review process in place—a review process that requires convincing preclinical data. However, even with a thorough review process, there is still the possibility of unanticipated harm. Indeed, the first somatic-cell gene-therapy patients, some of whom were children, could also have been harmed. One hopes that any harm caused to an individual through germ-line intervention will be reversible through the same sophisticated techniques that were employed to make the initial genetic change. If not, society will owe the injured subject or subjects both its gratitude and ongoing financial and social support. Nonetheless, in any sphere of innovative therapy, a first step must be taken at some point, and that first step will always be surrounded by some uncertainty.

If the early trials of germ-line intervention are undertaken when the preclinical research has matured, and with the aid of prior public review, I anticipate that few human beings will be harmed. If the technique is validated in humans and employed to combat serious genetic diseases like cystic fibrosis and sickle cell anemia, I also anticipate that all future generations will be grateful to both the researchers and the subjects who pioneered in the use of this promising technique.

NOTES

1. LeRoy Walters, "Editor's Introduction," *Journal of Medicine and Philosophy* 10 (August 1985): 211.

2. W. French Anderson, "Human Gene Therapy," *Science* 256 (May 8, 1992): 808–813.

3. "ADA Human Gene Therapy Clinical Protocol," *Human Gene Therapy* 1 (Fall 1990): 327–329, 331–362.

4. Edward M. Berger and Bernard M. Gert, "The Ethical Status of Germ-Line Therapy," *Journal of Medicine and Philosophy* 16 (December 1991): 676–679; Burke K. Zimmerman, "Human Germ-Line Therapy: The Case for Its Development and Use," *Journal of Medicine and Philosophy* 16 (December 1991): 600–603.

5. Jon W. Gordon, "Transgenic Animals," *International Review of Cytology* 115 (1989): 171–229; LeRoy Walters, "Human Gene Therapy: Ethics and Public Policy," *Human Gene Therapy* 2 (Summer 1991): 118.

6. Zimmerman, "Human Germ-Line Therapy," p. 595.

7. Eric T. Jeungst, "Germ-Line Gene Therapy: Back to Basics," *Journal of Medicine and Philosophy* 16 (December 1991): 589–590; Zimmerman, "Human Germ-Line Therapy," pp. 596–598.

8. Eric T. Juengst, ed., "Human Germ-Line Engineering," [thematic issue], *Journal of Medicine and Philosophy* 16 (December 1991): 587–694.

9. Juengst, "Germ-Line Gene Therapy," p. 590; Zimmerman, "Human Germ-Line Therapy," pp. 604–609; Alex Mauron and Jean-Marie Thévoz, "Germ-Line Engineering: A Few European Voices," *Journal of Medicine and Philosophy* 16 (December 1991): 652–663.

Commentary: Diversity

Susan C. Lawrence

As one of my ongoing research interests is the construction of "the" human body and how it has been gendered in anatomy texts, I am particularly concerned about how the subtle stereotypes that permeate our culture appear in scientific work. For me, one of the crucial ways that cultural assumptions enter into and shape science and medicine is through common language and conventional habits. As David Hull aptly said, those of us who point to questionable uses of language seem "overly picky" at times. Even worse, we can seem humorless, deliberately difficult, or "politically correct." Yet what appears trivial to those who use various terms and locutions, ones that have been acceptable for years, can be of major importance to those who *read* or *hear* them. I wish to highlight but one of the many implications that I see emerging from Hull's and Hubbard's challenging discussions of differences and how we think and talk about them in conjunction with the Human Genome Project (HGP). My remarks center on the connections between epistemology, ontology, and language, not to make a highly theoretical point, but to ask us—scientists and humanists—to change.

David Hull has noted that Linnaeus, a preeminent eighteenth-century systematist, has been designated the type specimen for *Homo sapiens* in the HGP. Hull criticizes this gesture as problematic primarily because it perpetuates the notion—however unintended by experienced biologists—that a type specimen captures some meaningful essence, or set of characteristics that define a species. Hull then notes that since Linnaeus hardly represents the majority of human beings anyway, he is a lousy type specimen even considering that type specimens are basically arbitrary representatives that taxonomists use. Biologists know quite well that the type specimen has more pragmatic than theoretical meaning. Why, then, name the HGP "specimen" after Linnaeus? This decision was perhaps but a well-intentioned effort to honor Linnaeus, a mere token of esteem. Yet it is hardly socially or culturally benign. It attaches the image of a short, northern European, Caucasian male to sequences of genetic information drawn from a large array of human beings of diverse racial and ethnic heritages, not to mention from women.

Ruth Hubbard's essay helps us to understand why choosing Linnaeus is not simply poor science or even merely bad taste. Setting him up as a type specimen is crammed full of meaning about how language embeds social assumptions and values into the supposedly neutral scientific process. All the goodwill in the world will not detach what geneticists say and do from a cultural context. That is precisely what Hubbard urges us to work on. If we can, on the one hand, all nod and agree that a short, white, blue-eyed male is not a representative human—then why do we (using "we" to refer to our culture) do it? If naming the type specimen "Linnaeus" is biologically arbitrary, scientifically questionable, and socially insulting—what possible purposes does it serve?

Hubbard has clearly emphasized how ideology and biology have intertwined. They continue to do so. Biological facts support political and social agendas. At the same time, cultural beliefs shape the ways that human beings produce, talk about, and envision those very "facts," particularly "facts" about "race" and gender, or other kinds of diversity, including disease and disability. The conference papers and discussion have underscored how we, as humanists and scientists, are very much entangled in categories, including normative categories in science, analytic categories of equal opportunity, and social categories of disabilities. Of course we must continue to use the categories developed to make some sense of biological, social, and political phenomena. But only if we constantly press for ways to have the presumptions underlying our theoretical and pragmatic definitions come into contact and conflict, will we have chances to keep up with the mutability of categories in different disciplines.

Race, as Hubbard stated and Hull implied, has no meaning as a biological concept—but it is constantly reaffirmed as a biological "fact" in the way that scientists make reference to it. Certainly "the" human genome map being constructed has no race—or does it? How will we talk about its potential diversity? At one level, the HGP reminds us that racial markers, especially for skin color, are trivial differences in genetic material compared with the extent of genetic similarity in humans. At another level, however, the HGP has been cast in terms of a universality that makes the promise to take diversity into account highly questionable.

Hull and Hubbard do agree that sex, unlike race, has biological meaning. Yet here, the cultural desire to keep sex dichotomized—male or female—continuously works to emphasize difference, not similarity. Resistance to blurring sexual distinctions—genetically, embryologically, developmentally—like the resistance to blurring sexual preference—does not come pristine from "nature." As many researchers have shown in various areas, the male-female polarity is such a cultural given that to challenge it means that we do indeed need to change the way we talk and write. Looking for "*the* sex gene," to elaborate on one of Hubbard's examples, places this fictional sequence on the same concrete, discrete level as "the" gene responsible for sickle cell anemia or cystic fibrosis or (as previously believed) success on IQ tests. Sexual differentiation (like most other characteristics), Hubbard reminds us, is not and never will be reducible to a single sequence. In my experience, scientists, medical practitioners, and humanists agree immediately and wholeheartedly that sex and gender differences have extraordinarily complicated origins. But far too many go on without pause to use the familiar, convenient language typified by "*the* sex gene," whether in coffee-room conversations or in scientific papers.

We maintain a habit of using universalizing terms (especially "the") in part because science (and other forms of academic knowledge) has a great stake in making claims that transcend individual variations, as Hull emphasized in his discussion of "species." Our language, our styles of speaking and writing, depend upon "essences," whether overtly fictional, merely conventional, or taken literally as empirical observations. Scholars have written for millennia on the problems of making universal statements and how they are (or are not) connected to "reality." This observation is not new. My concern is that I, as a conference participant or reader, simply cannot *tell* what stance a scientist is taking about the "reality" of her or his claims when he or she uses a locution like "*the* human genome" or "*the* sex gene." These phrases *sound* essentialist. They look as though scientists are making specific, meaningful, and significant claims about reality. Scientists, having a complex understanding of the work they are doing, might not

much care when they use "the human genome" literally, convention-
ally, or metaphorically. But then they should not be surprised at the
confusion, misunderstandings, and criticisms that subsequently arise.

My reflections upon Hull's and Hubbard's essays thus end with a
plea for accuracy and accountability. Stop saying the HGP will result
in a map of "the" human genome. Say, instead, that we are looking for
a composite map of *a* human genome constructed, for example, from
fragments of John Smith's chromosome 20 and Jane Doe's X chromosome. Stop looking for *the* sex gene, and look for *a* genetic sequence
involved in the differentiation of testes, as Hubbard suggests. To ease
the constant polarizing of discussion into nature versus nurture, into
"the" cause of a particular characteristic or behavior, we need to stop
creating these dichotomies and these ontological bits as we speak. To
shift away from "the" human genome, when appropriate, will, I suggest, do a great deal to display what the HGP will in fact produce,
including the extent and limitations on the diversity it encompasses.

Certainly we must use general terms—I am not naive enough to
suggest that perfect knowledge would follow from some sort of per-
fectly neutral and pseudo-objective vocabulary. What I offer as a
springboard for discussion is a plea that biologists be as precise about
syntax as they are about experimental design. If the confusions about
the HGP apparent at this conference rest on "mere" language, then
science will trundle on its way unchanged by due attention to the
specificity of the human material being studied. If, as I believe that
Hubbard has demonstrated, our "common" language provides ways
for assumptions about race, gender, and other differences to configure
science, then we will learn much more as we go along. At the very
least, we will be materially and conceptually clearer.

Commentary on Hull and Hubbard

David Magnus

In their essays, David Hull and Ruth Hubbard raise interesting criticisms of human genetics in general and the Human Genome Project (HGP) in particular. Hull has tried to show that the HGP is irrelevant to the issues that evolutionary biologists care about regarding species, races, and sexual dimorphism. He also raises an interesting objection to the HGP. Hull points out that the essence of species is the lack of an essence—that is, variation, both at a time and through time, is a prominent feature of all species, and genetic heterogeneity is crucial to the survival of any species. Hull claims that the HGP fails to take the lessons of evolutionary biology to heart. Embodied in talk about *the* human genome is a kind of essentialist view of species that is untenable. In sum, there is too much variation in any species to make the idea of a "typical member" (whose DNA is to be sequenced) more than a confusion.

Hubbard criticizes recent biological work which attempts to ground differences between socially or politically significant groups of people (races, genders) on genetic differences. She claims that such attempts at grounding most differences among humans in genetics is bound to fail because "human beings (*Homo sapiens*) are genetically a relatively homogeneous species." There is simply too little genetic diversity among humans for programmatic reduction of socially significant differences to genetic differences. It seems as if there is a significant difference of opinion here between the two over the amount of variation in our species. Hull claims that there is a lot of variation (in any species) and Hubbard claims we are genetically rather homogeneous.

Both views have a great deal of appeal. The social dangers of taking a particular type of person as the standard (or worse, the ideal) of humanity is frightening. But no less frightening is the use of genetics to give a false justification of the preservation of current social circumstances and power relations. Is there a way of reconciling the two views? I believe so. Is a glass which is half full, a lot of water or not enough? The appropriate answer depends on what one wants to do with it. It is more than enough to swallow a pill, but not enough to put out a fire. Similarly, whether our species is homogeneous or heterogeneous depends on what one does with the differences and similarities. Hubbard claims that about 75 percent of known genes are the same in all humans. But what is even more significant is that "because of the extent of interbreeding that has happened among human populations over time, our genetic diversity is pretty evenly distributed over the entire species."

Hubbard's argument requires only that the distribution of genetic differences that do exist do not fall into discrete groups which correspond to socially significant groupings, such as races. That could still leave ample variation in our species, from the point of view of the evolutionary biologist, perhaps enough to make problematic the notion of *the* human genome. Thus, while there is certainly a tension between the views expressed here, it is possible that there is ultimate compatibility, as long as the amount of variation in humans is sufficient to thwart the notion of a typical human, but not so great as to make it unproblematic that differences in genetic makeup are responsible for most socially significant traits.

I would now like to raise another concern over the HGP, based on some historical work that I have done in a very different context. If one looks carefully at developments in American biology at the turn of the century, one discovers a shift in emphasis taking place that has been characterized in a variety of ways, perhaps most popularly as a "revolt from morphology."[1] Typical stories recount the triumph of an experimental, quantitative, and reductive biology over a qualitative, speculative natural history, eventually giving rise to genetics.

The stories are wrong at a fundamental level. In fact, as several historians of biology have pointed out, the deep differences among groups of biologists often meant that there was no overlap in the sorts of problems being tackled, making comparisons between competing ways of doing science impossible.[2] Nonetheless, there was one significant area of overlap, and historians of biology, notably Garland Allen, have leaped on it as especially significant for understanding the triumph of the more experimental and reductive methodology. The historical details of the situation are rather surprising. Traditional natural historians followed what they called the Darwinian method of searching for consilience of inductions, that is, a variety of kinds of evidence pointing in the same direction. In contrast, experimentalists wanted to eliminate kinds of evidence which were not replicable—hence experimental evidence alone was relevant to science.

This led to a direct conflict over the mechanisms involved in speciation. A variety of data—biogeographic, geological, taxonomic, ecological—and in a wide range of organisms seemed to indicate that isolation was essential to speciation. However, on the basis of breeding experiments with *Oenothera*, Hugo de Vries concluded that evolution proceeded by macromutation with isolation and natural selection playing little if any role in evolution. The fact that there was experimental evidence in favor of de Vries's theory was sufficient for a growing number of biologists, in spite of the bulk of nonexperimental evidence that indicated the central importance of isolation to evolution. In the course of debates over this issue, prominent biologists argued that the *only* sort of evidence relevant to settling scientific issues is replicable, experimental evidence. As C. S. Gager, a prominent botanist, said:

> It [the mutation theory] is founded almost entirely upon experiment, and can be verified only by the same means. . . . For mere opinion and inference, and a priori impressions and prejudices, and inductions from field studies and comparative morphology there is absolutely no place. . . . If he [a critic] doubts that they [de Vries's theories] represent a general truth, a fundamental principle in biology, then let him await the fullness of time, for it is by repeated experiment . . . and by experiment alone that the general application must stand or fall.[3]

D. T. MacDougal, perhaps the leading champion of the mutation theory in this country, likewise rejected all nonexperimental evidence:

> The basal and underlying fault for misunderstandings consists in the fact that taxonomic and geographic methods are not in themselves, or conjointly, adequate for the analysis of genetic problems [genetic in the sense of origins]. The inventor did not reach the solution of the problem of construction of a typesetting machine by studying the structure of printed pages, but by actual experimenta-

tion with mechanisms, using printed pages only as a record of his success. Likewise no amount of consideration of fossils, herbarium specimens, dried skins, or skulls of fishes in alcohol may give actual proof as to the mechanism and action of heredity in transmitting qualities and characters from generation to generation.[4]

The naturalists stubbornly fought this battle in the early decades of this century, and they came to triumph in terms of our theoretical understanding of evolution. But, while they won the theoretical battle, they lost the methodological or epistemological war over how to do good science. Biology became increasingly experimental and reductive, in spite of the fact that it was less successful in the only context where a reasonable comparison could be made between the competing methods. This has resulted in a decline in research into areas and problems that once held prominence—areas such as systematics. The recent enthusiasm for molecular comparisons in systematics at the expense of traditional methods is a more recent example of this trend.

What is the relevance of all this for the HGP? If Hull is right, the HGP does not really have very much to say about the sorts of things which concern evolutionary biologists. But, if the epistemological trend I have described continues, one of two things is likely to happen. First, these issues will no longer seem as important, as money and prestige will be found elsewhere. Second, biologists will "force" the relevance of the HGP to these issues, to the detriment of good science.[5] The history of biology tells us that enthusiasm for experimental and reductive approaches—such as the HGP—will gain support even at the cost of other worthwhile approaches to problems. So the concern I raise is that we not forsake the problems of biology that do not easily admit of experimental or reductive solutions. Most biologists would undoubtedly agree that other areas of biology are important. But such pluralistic claims ring hollow when the practical realities result in positions, funding, and prestige going to some areas and not others.

David Starr Jordan, a leading naturalist and champion of the role of isolation in evolution in the 1920s, in describing biology made what he sometimes called "a plea for old-fashioned natural history":

> There is a recent tendency with many biologists to deprecate Darwin and his method of approach. . . . Discoveries connected with the physical basis of heredity and [subsequently] of Mendelian processes in variation and crossing have given an immense impetus to

the study of Genetics. . . . Moreover, as no one man can compass and weigh all kinds of evidence, that derived from field study, species-study, and taxonomy is being overlooked by workers in other fruitful fields. The basal truth remains: the method of Darwin of considering all relevant discovered fact is the only way to the acquisition of sound knowledge.[6]

Like Jordan, I hope our enthusiasm for the latest experimental approaches does not prevent or replace work on problems which do not easily fit into the experimental mold.

NOTES

1. Garland Allen, *Life Science in the Twentieth Century* (Cambridge: Cambridge University Press, 1975).

2. Jane Maienschein, Ronald Rainger, and Keith Benson, "Introduction: Were American Morphologists in Revolt?" *Journal of the History of Biology* 14 (1981): 83–87; Maienschein, "Shifting Assumptions in American Biology: Embryology, 1890–1910," *Journal of the History of Biology* 14 (1981): 89–113.

3. C. S. Gager, "De Vries and His Critics," *Science* 24 (1906): 89.

4. D. T. MacDougal, "Discontinuous Variation in Pedigree-Cultures," *Popular Science Monthly* 69 (1906): 210.

5. At the Center for Human Genome Studies Larry Deaven has recently suggested that the project will provide increasing support for the mapping and sequencing of model organisms, which could lead to evolutionary insights, particularly the evolution of the genome. This could be interpreted either as an example of forcing the relevance of the project to evolutionary concerns, or as an argument against Hull's claim that the HGP is irrelevant to evolutionary biology.

6. David Starr Jordan, unpublished manuscript, Stanford University Archives.

Response to Walters

John P. Boyle

My remarks will be confined to the ethical problems of germ-line intervention raised by Professor Walters. I invite you to think with me about the special problems that attend this particular kind of genetic engineering.

An interesting challenge to ethicists is to look for proper analogies to new problems in generally accepted analyses of already familiar ethical questions. I will propose some examples of actions with ethical implications, one old and one more recent, to see whether they might offer us some ethical insights as we deliberate about the problems and the promise of germ-line intervention.

I will assume in this brief exercise that standards of research safety, including preliminary theoretical work and work with animals have been met.[1]

My first example is old enough that I trust it will not seem tendentious—though it may seem far-fetched.

One of the great power rivalries of the ancient Mediterranean world was that between Rome and Carthage. That rivalry came to a head in the series of Punic Wars, the last of which began in 149 B.C.E. It was at that time that the Roman senator Cato the Elder ended his speeches in the Senate with the slogan *Carthago delenda est*, "Carthage must be destroyed." In 146 B.C.E. Rome overran Carthage, leveled the city, dispersed the population—and then, to ensure its victory, poured salt over the ruins to render the site of Carthage uninhabitable.

Now granted that war is hell, and granted that vengeance upon enemies is never a pretty business, I want to ask whether the pouring of salt over the ruins of Carthage did not add a special malice to the moral evils of killing and destruction. If the answer is affirmative, I want to ask, why?

It seems to me that the answer clearly is yes, the salting of Carthage was an added evil. If we ask why the added insult made a moral difference, it seems to me that the difference lies in the dimension of *time* that the salting added to the destruction of Carthage. By using salt to render the land incapable of supporting life for a longer time than anyone could see or imagine, the Romans did their terrible best to make the destruction of their enemy irreversible. The later generations which rebuilt a city named Carthage had to find another place for it.

One of the moral issues in germ-line intervention is time. While somatic-cell interventions can make possible the replacement of defective genes in one human being, the correction cannot affect later generations as germ-line interventions would. This time consideration applies both to potential benefits and to potential harms passed to subsequent generations.

My second example is recent. Early in 1991, when the war in the Persian Gulf was going badly for Iraq, we heard reports that Iraqi troops in Kuwait were setting fire to Kuwaiti oil wells—hundreds of them. We also heard of giant oil slicks being created by the release of oil into the Persian Gulf by the Iraqis. The destruction of the oil wells and the damage to shipping installations would have been a heavy blow to the economy of Kuwait in any event. But there were fears expressed at the time of the fires and the oil slicks that not only was the ecosystem of the Persian Gulf in danger, with potentially catastrophic consequences for the whole area, but the dense smoke from the hundreds of oil fires threatened to obscure the sun in areas downwind from Kuwait, including the Indian subcontinent. It was feared that the reduced sunlight would produce climatic changes which would have a devastating effect on crops in areas with enormous populations and barely adequate food production.

If the dimension of *time* aggravated the moral seriousness of the Roman destruction of Carthage, the dimension of *space* over which an imminent ecological disaster created by human beings threatened to spread seemed to many observers to aggravate in a similar way the moral seriousness of Iraq's actions in Kuwait. To be sure, the worst fears were not realized; the fires were put out much more quickly than expected and the oil slicks were not as destructive as had been feared. Yet many have judged that what Iraq did in creating a potentially disastrous situation for a vast area and for an incalculable period of time was morally far more serious than are evils affecting only a very limited time and space.

Moral deliberation about germ-line interventions suggests analogous concerns. It seems to me that one of our major moral concerns about germ-line interventions—both in the good that they might do and in the harm they might cause—arises from their potential for perdurance in time and expansion in space, especially over several generations. No wonder that the possibility of mistakes in such interventions weighs so heavily in any risk-benefit analysis.

I believe that there is also another consideration, though it is harder to pin down and surveys indicate that it may even be overridden in the minds of many people by the potential benefits of germ-line intervention. It is this: Even those who do not believe in God may find themselves opposed to "playing God" with the human genetic endowment. Although surveys show more people approve of germ-line intervention to correct genetic defects than oppose it, there remains a substantial minority who do oppose such interventions. Among such people there is, I suspect, a reluctance to tamper with a given in our nature even when the intervention is planned for reasons that are therapeutic and not eugenic.

We can gain some insight into the reasons for this judgment if the consequences of germ-line intervention are imagined as a stone dropping into a pond, the shoreline of which we cannot see. The stone striking the water sends ripples in every direction, beyond our ability to follow and measure them. The inevitable element of uncertainty about the consequences of our actions understandably can give rise to the judgment that there is a kind of sacral and therefore untouchable quality about the human genetic endowment. That judgment can in turn reinforce a firm utilitarian judgment that the risks of germ-line intervention simply are not worth the promised benefits.

Harsh experience has taught us to take such suspicions seriously. In places like Iowa, the advertising on the evening news is not by airlines pushing vacation trips to Florida or abroad, but from companies pushing everything from herbicides and pesticides to hybrid corn and soybeans or even the "boar power" that a newly tested crossbred animal promises to livestock raisers. Despite the benefits which scientific and technological advances have brought to agriculture and other fields, our generation is only too aware of the threats created, for example, by DDT and nitrates that accumulate in Iowa City's drinking water from the runoff of our fields and periodically make it necessary to warn pregnant women and mothers of small children that they should not drink the local tap water or give it to their children.

The injunctions to farmers familiar to radio and TV listeners in the Midwest that a farm chemical is labeled for restricted use and should be used only according to the directions on the label is a useful reminder that "doing no harm" is a more complex moral injunction than we may think, especially if considerations of time and space and human fallibility are factored in. Even if the human genetic endowment is not sacred in any proper sense, we can learn from the Romans and even from Saddam Hussein not that we should do nothing, but that we should make decisions with a lively sense of the moral responsibility we have to subsequent generations and with an equally lively sense that our ability to follow clearly the consequences of our actions very far in time or space is limited.

We have had enough experience over time and space and enough experience of the limits of our foresight to make us wary of "expert" judgments about the potential risks and benefits of many advances in science and technology. The caution with which public policy approaches germ-line interventions has strong moral warrants.

NOTE

1. Subcommittee on Human Gene Therapy, Recombinant DNA Advisory Committee, National Institutes of Health, "Points to Consider in the Design and Submission of Protocols for the Transfer of Recombinant DNA into the Genome of Human Subjects," *Human Gene Therapy* 1 (1990): 93–103.

Notes on Contributors

MITCHELL G. ASH, PH.D., is associate professor of history at the University of Iowa, where he teaches modern German history and history of science. He has written or coedited four books on the history of modern psychology. His current research topics include psychological twin research under Nazism, denazification of scientists after 1945, and émigré German-speaking scientists after 1933.

PETER DAVID BLANCK, PH.D., J.D., is professor of law at the University of Iowa College of Law. He is a senior fellow of the Annenberg Foundation Washington Program, exploring issues related to the implementation of the Americans with Disabilities Act. He serves as president of the American Association on Mental Retardation's Legal Process and Advocacy Division, and he is a member of the American Bar Association Commission on Mental and Physical Disability Law.

JOHN P. BOYLE, PH.D., is professor in the School of Religion at the University of Iowa. He is the author of the forthcoming *Teaching Authority in the Catholic Church: Historical and Theological Studies*, a former president of the Catholic Theological Society of America, and a senior director of the Society of Christian Ethics.

DAN W. BROCK, PH.D., is professor of philosophy and biomedical ethics as well as director of the Center for Biomedical Ethics in the School of Medicine at Brown University. He has published many papers in biomedical ethics as well as in moral and political philosophy and is the author of *Deciding for Others: The Ethics of Surrogate Decision Making* (with Allen Buchanan) and *Life and Death: Philosophical Essays in Biomedical Ethics*. He is currently at work on the Human Genome Project and the limits of ethical theory.

PANAYOT BUTCHVAROV, PH.D., is professor of philosophy at the University of Iowa. He is a past president of the American Philosophical Association, Central Division, editor of the *Journal of Philosophical Research*, and author of several books, the latest one being *Skepticism in Ethics*.

CHRISTINE CARNEY is a social work major at the University of Northern Iowa. She is currently in her internship at Trinity Regional Hospital, Fort Dodge, Iowa. She plans to pursue a master's degree in rehabilitation counseling.

WILLIAM E. CARROLL, PH.D., is professor of history at Cornell College. He is the editor of *Nature and Motion in the Middle Ages* and has published several articles in the history of science, especially concerning the relationship between religion and science.

DIANA FRITZ CATES, PH.D., is assistant professor of ethics in the School of Religion at the University of Iowa. She has published articles on various topics in ethics and moral psychology and is currently completing a book on the nature and value of compassion.

EVAN FALES, PH.D., is associate professor of philosophy at the University of Iowa. His publications include papers in the philosophy of science, philosophical semantics, and metaphysics and a book,

Causation and Universals. He is presently working on a book in epistemology and on topics in the philosophy of religion.

LARRY GOSTIN, J.D., is executive director, American Society of Law, Medicine and Ethics and visiting professor at the Georgetown University Law Center and the Johns Hopkins School of Hygiene and Public Health. He most recently served on the President's Task Force on National Health Care Reform.

RUTH HUBBARD, PH.D., is professor emerita of biology at Harvard University. In her laboratory research she has elucidated the molecular basis of light reception in vision. Since the mid-1970s, she has focused her attention on the sociology and history of biology with special emphasis on women's biology and health and on genetics. Her most recent books are *The Politics of Women's Biology* and *Exploding the Gene Myth* (with Elijah Wald).

DAVID L. HULL, PH.D., is Dressier Professor in the Humanities in the Department of Philosophy at Northwestern University. He has written and edited several books, including *Science as a Process*. He is currently working on ethical and social issues in the proper conduct of science.

KEVIN KOEPNICK, M.S., teaches biology and directs the science research program at City High

School in Iowa City. He has done research in geology, ecology, and molecular biology and has extensive experience in both pre-service and in-service teacher education. He has published papers in molecular biology and science education and is currently pursuing a Ph.D. in science education at the University of Iowa.

SUSAN C. LAWRENCE, PH.D., is assistant professor in the Department of History and medical historian in the College of Medicine at the University of Iowa. Her recent publications include "His and Hers: Depictions of Male and Female Anatomy in Anatomy Texts for Medical Students, 1890–1989" (with Kae Bendixen). Her current research concerns the historical relationships among human dissection, anatomy texts, and images in creating the "normal" body in western medicine.

DAVID MAGNUS, PH.D., is assistant professor of philosophy at Grinnell College. He is the author of several articles in the history and philosophy of biology, including the forthcoming "Down the Primrose Path: Competing Epistemologies in Early Twentieth Century Biology." He is currently working on a book in the history of species and speciation concepts.

ALAN I. MARCUS, PH.D., is director, Center for Historical Studies of Technology and Science;

director, Graduate Program in the History of Technology and Science; and professor of history at Iowa State University. He is the author of four books, most recently *Cancer from Beef: The DES Controversy, Federal Food Regulation and Consumer Confidence in Modern America*. He is currently working on a project entitled "the end of expertise in the age of individuation."

JOSEPH D. MCINERNEY, M.S., is director, Biological Sciences Curriculum Study, the Colorado College, Colorado Springs. In his seventeen years at BSCS, he has directed the development of numerous nationwide education programs, primarily in genetics and evolution. He is currently directing development of the second BSCS module on the Human Genome Project, to be distributed free of charge to all high school biology teachers in the United States.

DIANE B. PAUL, PH.D., is codirector, Program in Science, Technology, and Values, and professor of political science at the Boston campus of the University of Massachusetts. She has written on numerous topics pertaining to genetics and eugenics; a collection of these essays will be published as *Evolutionary Biology, Social History, and Political Power*. She is currently writing a nonspecialist book on the history of eugenics and human genetics and a longer scholarly monograph on the same subject.

KIMBERLY A. QUAID, PH.D., is director of the Predictive Testing Program for Huntington Disease and clinical assistant professor of medical and molecular genetics and psychiatry at the Indiana University School of Medicine. She has published many articles on predictive genetic testing and is active, both nationally and internationally, in developing protocols for testing. She is currently working on a casebook illustrating ethical issues arising from predictive genetic testing for late-onset disorders.

MICHAEL RUSE, PH.D., is professor, Department of Philosophy and Zoology, at the University of Guelph. He has edited and written several books, including *Taking Darwin Seriously* and *The Darwinian Paradigm*. He is currently completing a book on evolutionary naturalism.

LARRY THOMPSON, M.S., is a health science journalist at the *Washington Post*. He is currently writing a book on the Human Genome Project.

ELIZABETH J. THOMSON, M.S., R.N., is coordinator of genetics services research in the Ethical,

Legal, and Social Implications Branch of the National Center for Human Genome Research at the National Institutes of Health. She previously coordinated the statewide genetic counseling services for the University of Iowa from 1980 to 1992. She is coeditor of a forthcoming book on reproductive genetic testing and its impact on women.

LEROY WALTERS, PH.D., is professor of Christian ethics at the Kennedy Institute of Ethics and professor of philosophy at Georgetown University. He is coeditor with Tom L. Beauchamp of an anthology entitled *Contemporary Issues in Bioethics* and coeditor with Tamar Joy Kahn of an annual reference work entitled *Bibliography of Bioethics*. He and Julie Gage Palmer are writing a book on the ethics of human gene therapy, and he chairs the Recombinant DNA Advisory Committee at the National Institutes of Health.

ROBERT F. WEIR, PH.D., is director, Program in Biomedical Ethics, and professor in the Department of Pediatrics and the School of Religion at the University of Iowa. He has edited and written several books, including *Abating Treatment with Critically Ill Patients*. He is currently working on papers addressing ethical issues in genetics research.